# OXYGEN CHEMISTRY

# THE INTERNATIONAL SERIES OF
# MONOGRAPHS ON CHEMISTRY

# OXYGEN CHEMISTRY

Donald T. Sawyer

TEXAS A&M UNIVERSITY

Foreword by Professor R. J. P. Williams, FRS

New York   Oxford

OXFORD UNIVERSITY PRESS

1991

Oxford University Press

Oxford  New York  Toronto
Delhi  Bombay  Calcutta  Madras  Karachi
Petaling Jaya  Singapore  Hong Kong  Tokyo
Nairobi  Dar es Salaam  Cape Town
Melbourne  Auckland

and associated companies in
Berlin  Ibadan

Copyright © 1991 by Oxford University Press, Inc.

Published by Oxford University Press, Inc.,
200 Madison Avenue, New York, New York 10016

Oxford is a registered trademark of Oxford University Press

Library of Congress Cataloging-in-Publication Data
Sawyer, Donald T.
Oxygen chemistry / Donald T. Sawyer.
p.  cm. —
(The International series of monographs on chemistry ; 26)
Includes bibliographical references and index.
ISBN 0-19-505798-8
1. Oxygen. I. Title. II. Series.
QD181.O1S18 1991
546'.721—dc20  91-6766

2 4 6 8 9 7 5 3 1

Printed in the United States of America
on acid-free paper

# FOREWORD

Molecular dioxygen is a major constituent of the atmosphere and it is a striking fact that not only dioxygen but also ozone and water are not in equilibrium with either the inorganic or the organic materials on the surface of the earth, and here we include all living organisms. Life depends on the kinetic barriers to oxygen reactions. It is therefore outstandingly important for us to understand the kinetic constraints in the presence of the very strong thermodynamic drive which allows the controlled use of dioxygen as part of the food stock of aerobic life and as a transforming chemical in a series of reactions involving many minerals and man's materials. Here corrosion is only one aspect. There is also of course the opposite side of the problem which is the understanding of how the present out-of-balance atmosphere arose some $10^9$ years ago. Originally life did not have to protect itself from dioxygen, and obviously it could not use it in the present day fuel cells, but today life must also have protective devices against dangerous reduction products from dioxygen, for example superoxide and hydrogen peroxide. There is then in the on-going chemistry of the surface of the earth, in man's extensive industry from ore refining down to the most complex drugs, and in living processes the need to utilize dioxygen while controlling adverse factors in that utilization.

In this book Sawyer shows how all this chemistry can be managed and how much of it can be understood through careful consideration of intermediates. Not all of us will put the same value on this idiosynchratic valence-bond approach to the formation of oxygen compounds in Chapter 3. However this line of thinking has led Sawyer to look for and find reaction pathways, especially in non-aqueous media, which others have missed. It is then a challenge to the reader to reformulate, if he so desires, the multitude of facts about dioxygen collected together in the several chapters of this stimulating book. I think it is fair to say that the author knows he is being provocative and is writing from deep knowledge and experience. I know him well enough to be sure that he will welcome the challenge of others once they have sat and thought about the peculiarities of this most important of molecules, dioxygen. Through exchanges of information and views, especially in a stimulating manner, we can learn to appreciate the opportunities and problems presented by our unstable environment.

*University of Oxford*                                               R. J. P. Williams
*Oxford, England*
*March 1991*

# PREFACE

This book results from a conviction that oxygen is the most important element in the realm of chemistry. With the exclusion of the "chemically" uninteresting hydrocarbons, there is a greater diversity of oxygen-containing molecules than of carbon-containing molecules. In turn, the most important molecule is water, which is (a) the unique medium for biological chemistry and life, (b) the source of all of the $O_2$ in the atmosphere, and (c) the unique moderator of the earth's climate (large heat capacity from strong hydrogen bonding, and a rich vibrational absorption spectrum).

My interest in dioxygen began in 1957 after an introduction to the Clark Electrode by Dr. John Leonard of Beckman Instruments. With a Grant-in-Aid from the latter company we began a thirty-five year odyssey of the electrochemical reduction of dioxygen. The first chapter was accomplished by then sophomore Leonard Interrante at the University of California, Riverside (now Professor of Chemistry, Rensselaer Polytechnic Institution and Editor of *Chemistry of Materials*). His experiments established that dioxygen is reduced at metal electrodes by two pathways: (a) with passivated electrodes it is a metal-independent and pH-independent ECE mechanism, and (b) with metal electrodes that have freshly prepared surfaces there is a pre-chemical oxygenation of the metal electrode to give the redox thermodynamics of the metal oxide. This work was followed in 1965 by an investigation conducted by Professor Julian L. Roberts, Jr. (University of Redlands, Redlands, California) in which we discovered that dioxygen in aprotic solvents is reversibly reduced to superoxide ion ($O_2^{-\cdot}$). Subsequent research by a postdoctoral associate, Dr. E. T. Seo, Hughes Battery, Torrance, California) established that for all conditions the electron-transfer reduction of molecular oxygen is a one-electron process.

With the discovery that superoxide ion is a stable species in aprotic media, a series of investigations was initiated by Professor Margaret V. Merritt (Wellesley College) and Professor Roberts that characterized superoxide ion to be a strong Brønsted base and a powerful nucleophile.

During the 1980s another postdoctoral associate, Dr. Hiroshi Sugimoto, discovered that transition metals can activate hydrogen peroxide via their Lewis-acidity in base-free media for oxidase/dehydrogenase chemistry rather than Fenton chemistry. In an extension of this work a graduate student, Dr. Ceshing Sheu, discovered that the iron-picolinate complexes in a pyridine-acetic acid matrix activate hydrogen peroxide for the direct ketonization of methylenic carbons and the dioxygenation of arylolefins. A related iron(II) complex was found that activates dioxygen for the same chemistry. Current research by postdoctoral associates Hui-Chan Tung and Andrzej Sobkowiak has demonstrated that the $Co^{II}(bpy)_2^{2+}$ is able to affect similar activations of hydrogen peroxide. Other recent investigations by Dr. Ceshing Sheu and Dr. Andrzej Sobkowiak have led to the development of reaction mimics for the methane mono-oxygenase enzyme system (hydroxylase/reductase) and the cytochrome $P$-450 mono-oxygenase/reductase system) [based on the $Fe^{II}(DPAH)_2$ or $Co^{II}(bpy)_2^{2+}$, $O_2$, and PhNHNHPh combination].

In a more general sense I am grateful to all of my former graduate students and postdoctoral associates for their contributions to my understanding of oxygen chemistry. I am especially appreciative for useful and stimulating discussions with Professor R. J. P. Williams and Dr. H. A. O. Hill (University of Oxford), Professor Joan S. Valentine and Professor Christopher Foote (University of California, Los Angeles), Professor James Hurst (Oregon Graduate Institution), Dr. James Fee (Los Alamos National Laboratory), Professor Fred Basolo (Northwestern University), and Professor D. H. R. Barton and Professor Arthur E. Martell (Texas A&M University). Dr. Hill provided a detailed critique that was most valuable to the final editing.

My thanks to the administration of Texas A&M University for the award of a Study-Leave (July-December 1990), which made it possible to complete the manuscript at the University of Oxford. I thank Professor Malcolm Green for the kind hospitality of the Inorganic Chemistry Laboratory. I am grateful to the editor and the production staff of Oxford University Press, and especially for the cooperative approach provided by Ms. Louise C. Page, Assistant Editor. A special word of thanks to Mrs. Debora Shepard, who prepared the final camera-ready production manuscript. The support and encouragement of my wife, Shirley, has been essential to the completion of this endeavor.

*College Station, Tex.*                                                               D. T. S.
*March 1991*

# CONTENTS

# OXYGEN CHEMISTRY

# 1

# INTRODUCTION: WHY OXYGEN CHEMISTRY?

*$O_2$: Vital elixir from green plants beneath the sun*

## Chemistry: The science of molecules

The fundamental premise of chemistry is that all matter consists of molecules. The physical and chemical properties of matter are those of the constituent molecules, and the transformation of matter into different materials (compounds) is the result of their reactions to form new molecules. A molecule consists of two or more atoms held in a relatively fixed array via valence-electron orbital overlap (covalent bonds; chemical bonds).

In the nineteenth century chemists focused on the remarkable diversity of molecules produced by living organisms, which have in common the presence of tetravalent carbon atoms. As a result the unique versatility of carbon for the design and synthesis of new molecules was discovered, and the subdiscipline of organic chemistry (the science of carbon-containing molecules) has become the dominant part of the discipline. Clearly, the results from a focus on carbon-based chemistry have been immensely useful to science and to society.

Although most molecules in biological systems [and produced by living organisms (particularly aerobic systems)] contain oxygen atoms as well as carbon and hydrogen (e.g., proteins, nucleic acids, carbohydrates, lipids, hormones, and vitamins), there has been a long tradition in all of chemistry to treat oxygen atoms as "neutral counterweights" for the "important," character-determining elements (C, H, Al, Si, Fe, I) of the molecule. Thus, chemists have tended to take the most important element (oxygen) for granted. The chemistry curriculum devotes one or two year-courses to the chemistry of carbon ("Organic Chemistry"), but only a brief chapter on oxygen is included in the first-year and the inorganic courses. However, if the multitude of hydrocarbon molecules is from the incorporation of oxygen atoms in single-carbon molecules argues against the assignment of a "neutral character" for oxygen atoms [e.g., $C_n$(graphite), $CH_4$(g), $CH_3OH$(l), $CH_2(O)$(l), $HC(O)OH$(l), $(HO)_2C(O)$(aq), $CO$(g), $CO_2$(g)]. Just as the focus of nineteenth century chemists on carbon-containing molecules has produced revolutionary advances in chemical understanding, and yielded the technology to synthesize and produce useful chemicals, polymers, and medicinals; I believe that a similar focus on oxygen chemistry is appropriate and will have analogous rewards for chemistry, biochemistry, and the chemical-process technologies.

3

**The central role of oxygen in chemistry**

Oxygen atoms react spontaneously with the atoms of all elements except the noble gases to form polyatomic molecules via covalent bonds (shared valence electrons; see Chapter 3). Table 1-1 provides illustrative examples of such chemical transformations for hydrogen, nonmetals, metals, and transition metals.

*Water*

The HOH molecule (oxygenated hydrogen) is the most important oxygen-containing molecule and is fundamental to the realm of oxygen chemistry. It possesses exceptional ruggedness and thermal stability by virtue of strong chemical bonds

$$HOH(g) \xrightarrow{\Delta H_{DBE}, \ 119 \ kcal \ mol^{-1}} H\cdot(g) + HO\cdot(g) \qquad (1\text{-}1)$$

$$\text{(DBE; dissociative bond energy} \xrightarrow[\Delta H_{DBE}, \ 103 \ kcal \ mol^{-1}]{} H\cdot + \cdot O\cdot$$

Thus, 222 kcal mol$^{-1}$ must be transferred to HOH to produce atomic oxygen, which is a measure of the deactivation and stabilization of atomic oxygen via formation of a water molecule. Another important and unique characteristic of water molecules is their tendency to cluster via intermolecular hydrogen bonds (more properly hydrogen–oxygen bonds)

$$4\,HOH(g) \rightleftharpoons \begin{bmatrix} \begin{array}{c} H \\ | \\ H-\overset{..}{O}\text{:}H \\ H-\overset{..}{\underset{|}{O}}\text{:}H\text{:}\underset{|}{O}-H \\ H\text{:}\underset{|}{O}-H \end{array} \end{bmatrix} \qquad (1\text{-}2)$$

$$(0\text{--}100^\circ C)$$

In contrast, for ambient conditions $BH_3$, $CH_4$, $NH_3$, and HF are monomeric gas molecules (as are CO, $CO_2$, $H_2S$, and $SO_2$).

Liquid water also has exceptional solvation energies for small ions and thereby promotes the dissolution and heterolytic dissociation (ionization) of salt, acid, and base molecules [e.g., NaCl(s), HCl(g), and NaOH(s)].

$$NaCl(s) \xrightarrow{H_2O} Na(OH_2)^+_{4\text{-}6} + Cl(H_2O)^-_{4\text{-}6} \qquad (1\text{-}3)$$

$$\qquad\qquad\qquad [Na^+(aq)] \qquad [Cl^-(aq)]$$

This ability to interact strongly with positively and negatively charged ions is consistent with water's unique amphoteric (acid–base) properties, and is why it is fundamental to the Brønsted acid–base theory.

Table 1-1  Examples of the Chemical Transformations That Result When Elemental Molecules Are Oxygenated

| Element (state) | Oxygen derivatives |
|---|---|
| $H_2(g)$ | $HOH(l)$, $HOOH(l)$ |
| $(Li)_n(s)$ | $(Li_2O)_n(s)$ |
| $(Be)_n(s)$ | $(BeO)_n(s)$ |
| $(B)_n(s)$ | $(B_2O_3)_n(s)$ |
| $(C_4)_n$ (s, diamond) | $CO(g)$, $CO_2(g)$, $(HO)_2C(O)(aq)$, |
| $(C_6)_n$ (s, graphite) | $[C(O)OH]_2(s)$, $CH_3OH(l)$, |
| $CH_2(O)(l)$, | |
| $CH_4(g)$ | $HC(O)OH(l)$ |
| $N_2(g)$ | $N_2O(g)$, $\cdot NO(g)$, $\cdot NO_2(g)$, |
| $HON(O)_2(g)$, $NH_3(g)$ | $HON(O)(g)$, $NH_2OH$ |
| $O_2(g)$ | $O_3(g)$ |
| $Cl_2(g)$ | $Cl_2O(g)$ |
| $HCl(g)$ | $HOCl(aq)$, $HOCl(O)_2(aq)$, |
| $HOCl(O)_3(aq)$ | |
| $(Al)_n(s)$ | $(Al_2O_3)_n(s)$ |
| $(Si)_n(s)$ | $(SiO_2)_n(s)$ |
| $P_4(s)$ | $P_2O_3(s)$, $P_2O_5(s)$, $HOP(O)(aq)$, |
| | $(HO)_3P(O)(aq)$ |
| $S_8(s)$ | $SO_2(g)$, $SO_3(g)$, $(HO)_2S(O)_2(aq)$ |
| $(Mn)_n(s)$ | $(MnO)_n(s)$, $(MnO_2)_n(s)$, $Mn_2O_7(s)$ |
| $(Fe)_n(s)$ | $(FeO)_n(s)$, $(Fe_2O_3)_n(s)$, $(Fe_3O_4)_n(s)$ |
| $(Os)_n(s)$ | $(OsO)_n(s)$, $(Os_2O_3)_n(s)$, $(OsO_2)_n(s)$, |
| | $OsO_4(s \rightarrow g)$ |

$$HCl(g) + H_2O(l) \xrightarrow{\text{H}_2\text{O}} H_3O^+(aq) + Cl^-(aq) \qquad (1\text{-}4)$$

$$Na_2O(s) + H_2O(l) \xrightarrow{\text{H}_2\text{O}} 2\,Na^+(aq) + 2\,HO^-(aq) \qquad (1\text{-}5)$$

Although the water molecule is exceptionally rugged, it is a reductant (nucleophile) of strong oxidants (electrophiles),

$$Cl_2(aq) + 2\,H_2O(l) \underset{\text{H}_2\text{O}}{\rightleftharpoons} HOCl(aq) + H_3O^+(aq) + Cl^-(aq) \qquad (1\text{-}6)$$

$$SO_3(g) + 2\,H_2O(l) \underset{\text{H}_2\text{O}}{\rightleftharpoons} H_3O^+(aq) + HOS(O)_2O^-(aq) \qquad (1\text{-}7)$$

and an oxidant (electrophile) of strong reductants (nucleophiles)

$$2\,Na(s) + 2\,H_2O \longrightarrow H_2(g) + 2\,Na^+(aq) + 2\,HO^-(aq) \tag{1-8}$$

$$e^-(aq) + H_2O \longrightarrow H\cdot(g) + HO^-(aq) \tag{1-9}$$
$$\longrightarrow \tfrac{1}{2}\,H_2(g)$$

Water is the essential matrix for (C, N, O, H)-based life. Thus, all living biopolymers have a water–carbohydrate surface component via hydrogen–oxygen bonding. The inter- and intracellular aqueous phases are essential for the dissolution and transport of sugars, salts, nutrients, and some enzymes and cofactors for homogeneous, solution-phase metabolism and respiration. Life on Earth began in a reducing atmosphere without dioxygen or an ozone shield to filter short-wavelength solar radiation (capable of breaking C–C, C–N, and C–H bonds of biomolecules). Hence, primitive life began in the seas where liquid water provided an adequate filter for hard uv radiation.[1-3]

## Metals and non-metals

When the Earth first formed from a condensed cloud of gases and dust, those elements and molecules that are unstable in the presence of water reacted to form oxygenated products via the reduction of water to molecular hydrogen.

$$Ca(g) + H_2O(g) \longrightarrow H_2(g) + (CaO)_n(s) \tag{1-10}$$
$$\xrightarrow{\ H_2O(l)\ } Ca^{2+}(aq) + 2\,HO^-(aq)$$

$$2\,Al(g) + 3\,H_2O(g) \longrightarrow 3\,H_2(g) + (Al_2O_3)_n(s) \tag{1-11}$$

$$Fe(g) + H_2O(g) \longrightarrow H_{2(g)} + FeO(s) \tag{1-12}$$
$$\xrightarrow{\ H_2O(l)\ } Fe^{2+}(aq) + 2\,HO^-(aq)$$

$$2\,P(g) + 4\,H_2O(g) \longrightarrow 3\,H_{2(g)} + 2\,HOP(O)(g) \tag{1-13}$$

$$n\,Si(g) + 2n\,H_2O(g) \longrightarrow 2n\,H_2(g) + (SiO_2)_n(s) \tag{1-14}$$

The latter escaped from the atmosphere, but during its formation it reduced many elements and thereby provided a reducing environment at the surface of Earth.

$$H_2 + S \xrightarrow{\ \Delta\ } H_2S \tag{1-15}$$

$$H_2 + Cl_2 \xrightarrow{\Delta} 2\,HCl \tag{1-16}$$

$$2\,H_2 + C \xrightarrow{\Delta} CH_4 \tag{1-17}$$

$$3\,H_2 + N_2 \xrightarrow{\Delta} 2\,NH_3 \tag{1-18}$$

Table 1-1 provides other examples of oxygenated metals and nonmetals, with the most reduced forms probably present in the Earth's crust and oceans under a reducing environment. Most contemporary analyses[1-3] believe that the primitive atmosphere of Earth consisted of $N_2$, $H_2O$, $CO_2$, $CH_4$, $NH_3$, HCN, and $H_2S$. Hence, under such reducing conditions the transition metals were limited to their lowest oxide ($MO$ or $M_2O$), and the coinage metals (Cu, Ag, and Au) and platinum-group metals (Ru, Rh, Pd, Os, Ir, and Pt) were stable in the elemental metallic state. The transformation of the Earth's atmosphere to an oxidizing environment via the photosynthetic oxidation of water to dioxygen ($O_2$) (photosystem II of green plants) resulted in the autoxidation of the reduced forms of the molecules of the Earth's crust, oceans, and atmosphere (over a period of about 750 million years).

$$4\,FeO(s) + O_2 \longrightarrow 2\,Fe_2O_3(s) \tag{1-19}$$
$$\underset{FeO}{\big\lfloor} \longrightarrow Fe_3O_4(s)$$

$$2\,HOP(O)(aq) + O_2 \xrightarrow{H_2O} 2\,(HO)_3P(O)(aq) \tag{1-20}$$

$$2\,Cu(s) + O_2 \longrightarrow 2\,CuO(s) \tag{1-21}$$

$$2\,C + O_2 \longrightarrow 2\,CO \xrightarrow{O_2} 2\,CO_2 \tag{1-22}$$

$$4\,NH_3 + 3\,O_2 \longrightarrow 2\,N_2 + 6\,H_2O \tag{1-23}$$

Another unique characteristic of oxygen-containing molecules is their exceptionally strong $X$–$O$ bonds. For example, the gas-phase enthalpies for homolytic bond dissociation ($\Delta H_{DBE}$) are: H–O·, 103 kcal; HO–H, 119 kcal; CO, 257 kcal; BaO, 134 kcal; FeO, 93 kcal. Chapter 3 provides a more complete discussion of the chemical bonds for oxygen-containing molecules.

## Dioxygen

*Chemical history*

The discovery and isolation of molecular oxygen (dioxygen, $O_2$) was first reported in 1774 by Joseph Priestley, but was independently co-discovered by Scheele in Sweden, who gave it the name *fire air* (reported in 1777).[1] Because it did not support combustion in a manner equivalent to other fuels, Priestley named the substance "dephlogisticated air" ("air that does not burn"). He also discovered that green plants generate dioxygen. In 1777, Antoine L. Lavoisier identified dioxygen as a component of air, and named it *oxygen* on the basis that its reaction with other elements generated oxy acids (his alternative designation was *vital air*). He established that dioxygen is the oxidant of fuels in combustion, and that it is the essential oxidant for biological respiration. Because science was a hobby rather than a vocation with Lavoisier, his contributions were limited by available time and by a foreshortened life. Prior to the French Revolution, he made a living as a tax collector (always a hazardous occupation, but for him it resulted in the loss of his head by the guillotine). Scheele observed that pure dioxygen (1 atm) is toxic to plants; Priestley made similar observations.[1]

Thus, in the late eighteenth century Priestley, Scheele, and Lavoisier had established that dioxygen comprises about 21% (by volume) of the Earth's atmosphere, that it is essential for aerobic life and toxic to plant life at high concentrations, and that it is the essential oxidant for the combustion of organic molecules (fuels). Priestley also noted that nitric oxide reacts with $O_2$ to produce $NO_2$ and $N_2O_4$, which react with water to form $HON(O)$ and $HON(O)_2$ (nitrous acid and nitric acid).

About $2.5 \times 10^9$ years before dioxygen was chemically identified and characterized, it was added to the Earth's atmosphere as a natural product via photosynthetic oxidation of water by primitive green plants (blue-green algae). Life, which evolved on Earth in the absence of dioxygen some 3.5 billion years ago, was restricted to hydrogen sulfide as a source of hydrogen atoms for bacterial photosynthetic energy transfer.[2,3] With the development of blue-green algae and photosystem II some one billion years later, water replaced hydrogen sulfide as a source of reducing equivalents with dioxygen a byproduct. Prior to the development of these primitive marine plants, photosynthetic life was limited to bacteria that could avail themselves of hydrogen sulfide from marine sources. Although extensive production of dioxygen occurred with the appearance of blue-green algae, biological utilization of dioxygen (aerobic life and oxidative metabolism) had to await another 750 million years for significant quantities of dioxygen to become available. This amount of time was required to produce sufficient dioxygen to titrate the immense amount of reduced iron [Fe(II)] present in the Earth's oceans. Until this was precipitated as iron(III) oxide, the atmospheric concentration of oxygen was less than 1%. Thus, about 1.7 billion years ago there was an almost instant transformation of the atmosphere to 17-21% dioxygen.[2] Until then life was restricted to the aquatic environment of the ocean. The absence of atmospheric oxygen precluded the photosynthetic production of ozone, which is the essential shield from the solar ultraviolet radiation that is sufficiently energetic to break the chemical bonds of organic compounds.

*Biosynthesis of dioxygen*

Green-plant photosynthesis utilizes water as a source of reducing equivalents (hydrogen atoms).   As a result there is a parallel evolution of dioxygen from the plant cells (chloroplasts) with the photosynthetic "fixing" of carbon dioxide as carbohydrate.   The conventional wisdom for the mechanism of photosynthesis includes two independent photosystems that pump reducing equivalents from water (photosystem II) through a cascade of steps controlled by enzymes to a $CO_2$ reduction center (photosystem I).[4]   The essential components for the dehydrogenation of water include manganese, chloride ion, calcium ion, and bicarbonate ion.   Green plants that are deprived of manganese in the chloroplast cells of the leaves cease to evolve oxygen.   Full photosynthetic activity and oxygen evolution can be restored by the administration of manganese.

Photosystem II (PS II) within higher plants represents a solar-energy-driven process that removes hydrogen atoms from water to form molecular oxygen ($O_2$, dioxygen) via an overall four-electron transfer reaction.[5]

$$2\,H_2O \rightarrow O_2 + 4\,[H^+ + e^- \equiv H], \quad E^{o'}(pH\,5),\ +0.93\ V \text{ versus NHE} \qquad (1\text{-}24)$$

where NHE is the normal hydrogen electrode.   The mechanistic sequence that facilitates this reaction involves an initial photoactivation process in which light is trapped and converted to chemical energy.   This energy is subsequently transferred to other protein complexes within the lipoprotein membrane, where the chemical transformation of water to dioxygen occurs.

Most reviewers[5-8] now argue that photosynthetic oxygen evolution results from a sequential four-step electron-transfer process in which oxidizing equivalents from chl $a^{+\cdot}$ are accumulated in a "charge-storing" complex to accomplish the concerted four-electron oxidation of two $H_2O$ molecules to one $O_2$ molecule.   The photooxidant (chl $a^{+\cdot}$) and reductant (pheo$^{-\cdot}$) are one-electron transfer agents, and the matrix is the lipoprotein thylakoid membrane.   Hence, evaluation and consideration of the one-electron redox potentials for PS II components within a lipoprotein matrix are necessary in order to assess the thermodynamic feasibility of proposed mechanistic sequences.

There is still substantial debate as to the number of manganese atoms necessary to catalyze the transformation of water to dioxygen, but at least two appear to be essential, and possibly as many as six.   The valency changes and the arrangements of the metal atoms in this multinuclear cluster remains uncertain. However, within the thermodynamic constraints of the photosynthetic redox process, the green plant accomplishes a transformation that has not been duplicated by chemists in the laboratory.   Hence, the majority of contemporary models[5-8] for the charge-storing system of PS II include a polynuclear manganese complex as its functional component, with the progression through the redox states identified with sequential increases in the overall oxidation state (covalence) of the manganese centers.   This progress through redox states is classified by $S$ states, in which each higher $S$ state is a one-electron oxidation of a lower $S$ state.   The $S_4$ state evolves molecular $O_2$ ($S_4 \rightarrow S_0 + O_2$).

Thus, most of the recent experimental data have been rationalized in terms of an active site that contains binuclear- or tetranuclear-manganese clusters with coordinated hydroxide ligands.[7]   The majority of the mechanisms for dioxygen evolution that are based on such models involve (1) sequential oxidation of the metal centers, (2) electron transfer from the bound hydroxide ligands to produce adjacent "crypto-hydroxyl" metal centers, which couple to form a peroxide-bridged intermediate, (3) reoxidation of the metal centers by oxidized Z (proximal tyrosine radical), and (4) their intramolecular oxidation of the peroxide bridge to produce dioxygen and regenerate the manganese-catalyst.

## $O_2$, unique natural product; revolutionary transformation of biology and chemistry

The almost instantaneous increase in the concentration of dioxygen in the oceans had revolutionary consequences for the existent anaerobic life. The vast majority of species could not adapt rapidly enough and were destroyed (or retreated to an oxygen-free environment). Some species adapted and others evolved to oxygen-utilizing aerobes. With dioxygen present in the atmosphere, solar radiation transformed a small fraction of it to ozone and thereby provided a protective shield against short-wave-length uv light.

$$3\,O_2 \xrightarrow{\ h\nu\ } 2\,O_3 \tag{1-25}$$

In turn this made it possible for the newly evolved aerobes to leave the marine environment and establish terrestrial life, and the subsequent evolutionary development of higher forms of life. For aerobic life to develop and make effective use of dioxygen (as an energy source via oxidation of organic molecules) required the biosynthesis of a series of metalloproteins to facilitate the transport and the selective reactivity of dioxygen and its activated forms ($O_2^-\cdot$, HOO$\cdot$, HOOH, HOO$^-$, and HO$\cdot$) in respiration, metabolic, and detoxification processes. Although dioxygen is vital for biological oxidations and oxygenation reactions, the mechanisms and energetics for its reduction and for the interconversion of the various dioxygen species have been characterized to a reasonable degree only during the past two decades.

In the chemical realm the dioxygen molecule ($\cdot O_2\cdot$, triplet ground state) is a unique reagent with two unpaired electrons (biradical) and a bond order of two. With its introduction into the oceans of the planet as a biproduct of green-plant photosynthesis, the transformation of the reducing environment into an oxidized state began. Those transition metals that were present autoxidized [Fe(II) $\rightarrow$ Fe$_2$O$_3$(s), Fe$_3$O$_4$(s),; Mn(II) $\rightarrow$ MnO$_2$(s), etc.] and, in most cases, precipitated as metal oxides. The same was true for the reduced forms of organic molecules [MeCH(O) $\rightarrow$ MeC(O)OH, RSH $\rightarrow$ RSSR, H$_2$S $\rightarrow$ S$_8$, etc.].

Aerobic life makes use of dioxygen as an oxidant in all of its respiration and oxidative metabolic processes. These are processes whereby the reducing equivalents of food (carbohydrates, fats, and proteins) are oxidized by dioxygen to give water, carbon dioxide, and molecular nitrogen. Although the transformation is equivalent to combustion, biology controls (via biological catalysts known as *enzymes* and *cofactors*) the reaction pathway in specific steps that release and/or store energy (via production of intermediate products), whereas combustion is a free-radical autoxidation process that requires a high-temperature initiation step. For example, the efficient transformation of the oxidizing energy of dioxygen into heat and stored chemical energy is accomplished in eukariotic cells with a protein known as *cytochrome-c oxidase*. This system contains two iron–porphyrin (heme) groups and at least two copper atoms, and is able to transform about 80% of the redox energy stored in dioxygen (chemists and engineers have not been able to achieve a comparable efficiency with synthetic catalysts). Aerobic life also makes use of a vast array of metalloproteins for the selective catalysis of processes that produce useful biomolecules from the reaction of dioxygen and hydrogen peroxide with specific substrates.[9] In most cases, these biological catalysts include one or more transition metals (iron, copper, manganese, and molybdenum) at their active sites. Although oxidases and peroxidases accomplish their chemistry by hydrogen-atom transfer (or electron–proton transfer), there are other catalysts (oxygenases) that transform substrates by the addition of one or two oxygen atoms from the dioxygen molecule.

A limiting factor for the development of higher forms of life and their attendant greater energy fluxes (metabolic rates) was, and continues to be, the low solubility of dioxygen (in water with an $O_2$ partial pressure of one atmosphere its concentration is approximately one millimolar, whereas in the gas phase the concentration is approximately 45 millimolar). The development of warm-blooded animals that are capable of vigorous physical activity required systems to increase the solubility and the rate of transport for dioxygen to all parts of the organism. What has evolved is a series of iron and copper proteins that reversibly bind and transport dioxygen via the blood streams of animals and fishes. In the case of mammals these are the heme proteins known as *hemoglobin* and *myoglobin*, with marine crustaceans the equivalent system is *hemocyanin* (a binuclear copper-containing protein), and in marine worms oxygen transport is via *hemerythrin* (a binuclear non-heme-iron protein).

With the constructive utilization of dioxygen as an oxidant, the formation of reduced intermediate oxygen species is inevitable. Even in the case of cytochrome *c* oxidase, a nominal four-electron reducing agent that converts dioxygen to water, there is evidence that some of the $O_2$ so processed leaks from the system in the form of superoxide ion ($O_2^{-\cdot}$).[10] Likewise, a number of oxidases transform dioxygen to hydrogen peroxide (HOOH). Both $O_2^{-\cdot}$ and HOOH are toxic via their interaction with various biological molecules. To this end aerobic organisms have developed protective agents in the form of catalase, a heme protein that catalyzes the disproportionation of HOOH to $H_2O$ and $O_2$, a series of metal-containing proteins known as superoxide dismutases that likewise disproportionate $O_2^{-\cdot}$ to HOOH and $O_2$. The chemical reactivities and activation of HOOH and $O_2$ are discussed in Chapters 4 and 6, the chemistry of oxy radicals is presented in Chapter 5, and the reactivity of $O_2^{-\cdot}$ is summarized in Chapter 7.

*Activation and reactivity of dioxygen species ($\cdot O_2\cdot$, HOO$\cdot$, $O_2^{-\cdot}$, HOOH)*

Because the ground state of dioxygen is a triplet with two unpaired electrons with parallel spins, its reactivity with organic molecules in the singlet state is severely restricted, particularly at room temperature. Such a limited reactivity is undoubtedly essential for the product of a biosynthetic process (a product with high reactivity would have destroyed the "maker"). Although dioxygen is a strong oxidant at $pH$ 7 when utilized as a concerted four-electron transfer agent, it is an extremely weak one-electron oxidant [$O_2 + e^- \rightarrow O_2^-$; $E°$, -0.16 V versus NHE (unit molarity for standard states)] (see Chapter 2 for a full discussion of the redox chemistry of dioxygen species). When the low reactivity of dioxygen is coupled with its limited solubility in aqueous solutions, constructive utilization makes it necessary for nature to provide solubilization, transport, and activation systems. With these the organism can control when, where, and to what extent the redox energy within the dioxygen molecule is utilized for constructive purposes in a biochemical process. Thus, higher animals have developed reversible binding systems (such as hemoglobin) for the concentration and transport of dioxygen, and for the controlled catalysis of the constructive transformation of biological substrates to useful products and thermal energy.

The reduction of dioxygen in an inert matrix by electron transfer is illustrated (Fig. 2-5) and discussed in Chapter 2.[11] The cyclic voltammogram indicates that molecular oxygen is reduced by one electron in a reversible process to give a stable anion radical, which is reoxidized to dioxygen. This experiment confirms that the thermodynamically reversible redox potential for dioxygen in aprotic media is about -0.65 V versus NHE. The addition of an electron to dioxygen yields $O_2^-$·, which is a much stronger oxidizing agent than $O_2$, and a species that can abstract protons from weak acids to give a neutral radical (HOO·) that spontaneously decomposes to HOOH and $O_2$.

$$O_2 + e^- \xrightarrow[\;H_2O,\, pH\, 7\;]{E°', \text{ -0.16 V versus NHE}} O_2^- \cdot \xrightarrow[\;HA\;]{\bar{e},\ +0.89\text{ V}} HOOH + 2\ A^- \qquad (1\text{-}26)$$

$$\xrightarrow{HA} HOO\cdot + A\text{-}$$

$$\xrightarrow{HOO\cdot} HOOH + O_2$$

The shift in the redox potentials for $O_2$ within aprotic solvents [-0.65 V in dimethylformamide (DMF) versus -0.16 V in $H_2O$] is the result of a much weaker solvation of anions by such media relative to water (for $O_2^-$·, about 88 kcal mol$^{-1}$ in DMF versus 100 kcal mol$^{-1}$ in $H_2O$). Likewise, the addition of base to hydrogen peroxide results in its spontaneous decomposition.

$$2\ HOOH + HO^- \rightarrow O_2^- \cdot + H_2O + HO\cdot \qquad (1\text{-}27)$$

$$\longrightarrow O_2 + HO^-$$

Hence, the consequence of the addition of an electron (or hydrogen atom) to dioxygen is the production of finite fluxes of an entire group of reactive intermediates [including the strongest oxidant of the oxygen family, hydroxyl radical (HO·)].

$$O_2 + e^- \xrightarrow{H_2O, HA} \longrightarrow \longrightarrow (O_2^{-\cdot}, HOO\cdot, HOOH, HOO^-, HO\cdot, HO^-) \quad (1\text{-}28)$$

Because approximately 15% of the $O_2$ that is respired by mammals appears to go through the superoxide state, the biochemistry and reaction chemistry of the species is of interest to those concerned with oxygen toxicity, carcinogenesis, and aging. The most general and universal property of $O_2^{-\cdot}$ is its tendency to act as a strong Brønsted base.[12] Although the acidity of perhydroxyl radical (HOO·) in water ($pK_a$ 4.9) is the same as acetic acid, the strong proton affinity of $O_2^{-\cdot}$ is due to its redox-driven propensity to be $O_2$ and HOOH.

$$2\,O_2^{-\cdot} + H_2O \rightleftharpoons O_2 + HOO^- + HO^-, \quad K = 2.5 \times 10^8 \text{ atm} \quad (1\text{-}29)$$

Thus, two superoxide ions plus a water molecule are converted to two strong bases. This strong proton affinity manifests itself in any media and is exemplified by the ability of $O_2^{-\cdot}$ to deprotonate butanol.

In aprotic media, $O_2^{-\cdot}$ exhibits exceptional reactivity as a nucleophile towards aliphatic halogenated hydrocarbons and carbonyl carbons with adequate leading groups. Chapter 7 discusses the reaction chemistry of $O_2^{-\cdot}$ with electrophiles, reductants, transition metals, and radical species, and the chemistry of its conjugate acid (HOO·) is outlined in Chapter 5.

The preceding sections indicate that hydrogen peroxide (HOOH) is a natural intermediate in the oxidative metabolism of air-utilizing organisms. For example, it is the major product from the enzymatic oxidation of glucose and of bases such as xanthine. (The latter is oxidized to uric acid by molecular oxygen, which is reduced to HOOH).

The existence of proteins such as *superoxide dismutase* and *catalase* in aerobes often is cited in support of the conclusion that $O_2^{-\cdot}$ and HOOH are biological toxins. The present understanding of the biological utilization of hydrogen peroxide includes enzyme-facilitated processes that involve its disproportionation, the dehydrogenation of organic molecules, and the production of hypochlorous acid in neutrophiles.

$$HOOH + HOOH \xrightarrow{\text{catalase}} O_2 + 2\,H_2O \quad (1\text{-}30)$$

$$RCH_2OH + HOOH \xrightarrow{\text{peroxidase}} RCH(O) + H_2O \quad (1\text{-}31)$$

$$HCl + HOOH \xrightarrow{\text{myeloperoxidase}} HOCl + H_2O \quad (1\text{-}32)$$

Given that the *myeloperoxidase* enzyme within the neutrophils produces hypochlorous acid, an intriguing possibility is that it may combine with another hydrogen peroxide molecule. Such a reaction in the laboratory produces a stoichiometric quantity of singlet dioxygen ($^1O_2$), which is selectively reactive with conjugated polyenes.

Although the activation of hydrogen peroxide by reduced transition metals has been known for almost 100 years as Fenton chemistry,

$$HOOH + Fe^{II}L_2 \xrightarrow{\ pH\,2\ } L_2Fe^{III}\text{-OH} + HO\cdot \qquad k \sim 100\ M^{-1}\ s^{-1} \qquad (1\text{-}33)$$

$$Fe^{II}L_2 + HO\cdot \to L_2Fe^{III}\text{-OH} \qquad\qquad k \sim 2 \times 10^8\ M^{-1}\ s^{-1} \qquad (1\text{-}34)$$

$$RH + HO\cdot \to R\cdot + H_2O \qquad\qquad k \sim 10^8\text{-}10^{10}\ M^{-1}\ s^{-1} \qquad (1\text{-}35)$$

$$HOOH + HO\cdot \to HOO\cdot + H_2O \qquad\qquad k \sim 1 \times 10^7\ M^{-1}\ s^{-1} \qquad (1\text{-}36)$$

there continues to be controversy as to whether the reaction of Eq. 1-33 occurs within healthy aerobic organisms. (The chemistry of HO· and other oxy radicals is discussed in Chapter 5.) Because free reduced iron is necessary to initiate Fenton chemistry, its relevance to biology depends on the presence of soluble inorganic iron. The transition metals of metalloenzymes usually are buried within the protein matrix, which provides an immediate environment that is distinctly different from that of bulk water (with respect to dielectric constant, proton availability, ionic solvation, and reduction potential). We have argued that dipolar aprotic solvents such as dimethyl sulfoxide, acetonitrile, pyridine, and dimethylformamide (polypeptide monomer) more closely model the matrix of biological metals than does bulk water.[13-15]

To this end, studies of the metal-ion activation of hydrogen peroxide in aprotic media have been undertaken, and the results have been compared with those in an aqueous matrix for the same substrates. Whereas traditional Fenton chemistry in aqueous solutions produces a diverse group of products via the radical chemistry that is induced by HO·, the products in a nonbasic matrix such as acetonitrile are characteristic of dehydrogenations, mono-oxygenations, and dioxygenations.[13] That traditional Fenton chemistry does not occur is confirmed by the complete retention of the reduced oxidation state for iron. Activation is believed to be by the iron(II) center, which is a strong Lewis acid that weakens the O–O bond of hydrogen peroxide (HOOH). This induces a biradical nature for the HOOH adduct, and results in the two-electron oxidation of organic substrates without intermolecular electron transfer from the iron(II). Similar activation is observed when $Fe^{III}Cl_3$ is used with HOOH in anhydrous acetonitrile.[14] Again, the activation appears to be via the Lewis acidity of $Fe^{III}Cl_3$ whereby the HOOH acquires substantial oxene (O) character for direct biradical electrophilic insertion into the $\pi$ bond of the substrate olefin (stereospecifically in the case of norbornene) to give the epoxide.

$$Fe^{III}Cl_3 + HOOH \to \left[ Cl_3Fe\overset{V}{\underset{OH}{\overset{OH}{\diagup}}} \longleftrightarrow Cl_3Fe^V{=}O(OH_2)] \right] \qquad (1\text{-}37)$$

The activation and reactivity of HOOH and hydroperoxides are discussed in

Chapter 4.

*Cytochrome P-450* is a unique agent for the stimulated activation of dioxygen to the peroxide state with subsequent reaction with organic substrates. Three model substrate reactions often are used to characterize this important enzyme system.

$$RCH{=}CHR' + O_2 + 2\,e^- + 2\,H^+ \xrightarrow{\text{cyt } P\text{-}450} RCH\overset{O}{\overset{\diagup\diagdown}{-}}CHR' + H_2O \qquad (1\text{-}38)$$

$$PhN(CH_3)_2 + O_2 + 2\,e^- + 2H^+ \xrightarrow{\text{cyt } P\text{-}450} PhNHCH_3 + H_2C(O) + H_2O \qquad (1\text{-}39)$$

$$RCH(OH)CH_2OH + O_2 + 2\,e^- + 2\,H^+ \xrightarrow{\text{cyt } P\text{-}450} RCH(O) + H_2C(O) + 2\,H_2O$$
$$(1\text{-}40)$$

Cytochrome *P-450* also catalyzes the insertion of an oxygen atom in aromatic and aliphatic carbon–hydrogen bonds. Thus, in the case of benzene the primary product from reaction with activated oxygen is phenol. The functional goal is to solubilize foreign substances in order that they may be excreted. However, in this case, the solubilized substance is extremely toxic. In contrast, toluene is converted to benzoic acid by cytochrome *P-450*. Hence, the toxic response in human liver to the ingestion of benzene and toluene is substantially different.

On the basis of the arguments presented in Chapter 3 and a series of model studies (discussed in Chapters 4 and 6), the valence-electron bonding for the active forms of catalase, peroxidase, and cytochrome *P-450* are formulated with uncharged oxygen atoms (instead of the high-valent metal-oxo formulations that are in contemporary favor).

$$\text{Catalase:} \quad \underset{\text{(tyrosine)}}{PhO}{-}\overset{|}{\underset{|}{Fe}}{}^{V}{=}O \quad \text{(Compound I)} \qquad (1\text{-}41)$$

$$\text{Peroxidase:} \quad \underset{\text{(histidine)}}{\text{(imid)}}\,\overset{|+}{\underset{|}{Fe}}{}^{IV}{=}O \quad \begin{array}{l}\text{(Compound I, horseradish}\\ \text{peroxidase)}\end{array} \qquad (1\text{-}42)$$

$$\text{Cytochrome } P\text{-}450: \quad \underset{\text{(cysteine)}}{RS}{-}\overset{|}{\underset{|}{Fe}}{}^{V}{=}O \qquad (1\text{-}43)$$

Hence, their chemistry is characteristic of atomic oxygen [derived from HOOH or $(O_2 + 2\,H^+ + 2\,e^-)$ and transiently stabilized by the iron center]. The relative reactivity of the three reactive intermediates is inversely proportional to their respective Fe–O bond energies, which are primarily dependent on the electron density around the iron centers and their covalence (number of covalent bonds). This dimension of oxygen activation is presented in Chapter 3, and also is a part of Chapters 4 and 6.


*Utilization of dioxygen and hydrogen peroxide in biology*

Although the present volume is directed to the chemistry of oxygen-containing molecules, the fields of biology, biochemistry, and biophysics have a major interest in the utilization of dioxygen and hydrogen peroxide by aerobic organisms. Much of the older literature has treated the enzymology of $O_2$ and HOOH activation in nonmolecular terms, and categorized it as respiration, oxidative metabolism, and nutrition. However, during the past two decades there has been a dramatic change with a recognition that biological systems consist of molecules that are synthesized from other molecules (and are transformed into different molecules) via chemical reactions. The situation is complicated by the macromolecular nature of the crucially important enzymes (biological catalysts), proteins, nucleic acids, and biomembranes. Often the chemically active center of an enzyme (e.g., cytochrome *P*-450, a thiolated heme protein) is within a protein matrix that provides (1) structural constraints and rigidity, (2) a unique chemical environment, and (3) selective solution channels for substrate reactants. This complication has prompted chemists to study biological systems via the use small-molecule models of proteins [e.g., tetra-phenyl-porphyrinato-iron(II) in place of myoglobin]. In some cases this has proven effective, but for the dioxygen- and HOOH-activating metalloproteins (e.g., cytochrome-*c* oxidase and horseradish peroxidase) the protein matrix is an essential part of their chemistry. A complete chemical understanding of such biological systems will require studies and characterization of the natural proteins with an appreciation that the laws of chemistry are universal, including those for oxygen-containing molecules.

The present volume provides a concise discussion of the chemical characteristics of dioxygen species ($O_2$, $O_2^-\cdot$, HOO$\cdot$, and HOOH), and of their interactions with other molecules. However, it is beyond the scope of this work to discuss the chemistry of dioxygen species and proteins, which is within the traditions of biochemistry and biophysics. I urge all interested scholars to become familiar with the state of understanding of dioxygen and hydrogen peroxide utilization in biology. I hope many will be inclined to apply the principles of oxygen chemistry to further studies of dioxygen- and hydrogen peroxide–activating proteins (e.g., oxidases, peroxidases, mono-oxygenases, dioxygenases, ligninase, and penicillinase).


**Suggested readings on biological $O_2$ and HOOH utilization**

O. Hayaishi (ed.). *Molecular Mechanisms of Oxygen Activation.* New York: Academic Press, 1974.

J. V. Bannister and W. H. Bannister (eds.). *The Biology and Chemistry of Active Oxygen.* New York: Elsevier, 1984.

L. L. Ingraham and D. L. Meyer. *Biochemistry of Dioxygen.* New York: Plenum Press, 1985.

T. E. King, H. S. Mason, and M. Morrison (eds.). *Oxidases and Related Redox Systems.* New York: Alan R. Liss, Inc., 1988.

T. Vanngard (ed.). *Biophysical Chemistry of Dioxygen in Respiration and Photosynthesis.* Cambridge, U. K.: Cambridge University Press, 1988.

## Process chemistry that utilizes $O_2$, HOOH, and $O_3$

Although the introduction of $O_2$ into the Earth's atmosphere revolutionized biology and gave birth to oxidative metabolism and the associated biochemistry [oxidases, peroxidases, superoxide dismutases, mono-oxygenases, dioxygenases, lipid peroxidation and autoxidation, and antioxidants (ascorbic acid, α-tocopherol, glutathione, and thiols)], the chemistry of dioxygen and hydrogen peroxide has not enjoyed the attention of most chemists. However, process engineers appreciate an essentially free reagent (air) and have developed major processes that are based on the catalytic activation of $O_2$:

1. Sulfuric production from $H_2S$ and $SO_2$ with $V_2O_5(s)$ as the catalyst;

2. Ethylene oxide production from ethylene with supported Ag(s) as the catalyst;

3. Cyclohexanol and cyclohexanone from cyclohexane with supported Co(s) as the catalyst.

Likewise, hydrogen peroxide, perborates, and peracids are used as bleaches and decolorizing agents, disinfectants, and general oxidizing agents. There is increasing interest in the use of ozone ($O_3$) for water purification (in place of chlorine), and as a clean oxygenating reagent. Nature for the past 1.7 billion years has devised highly selective catalysts for the constructive use of dioxygen and hydrogen peroxide to synthesize molecules. Through an understanding of the chemistry for these biological oxygen-activation processes new catalyst systems can be designed. Several examples of new homogeneous HOOH-activating and $O_2$-activating metal complexes are described in Chapters 4 and 6, and illustrate the potential to duplicate and, in some cases, to surpass Nature via a detailed knowledge of oxygen chemistry.

## References

1. Gilbert, D. L. (ed.). *Oxygen and Living Processes. An Interdisciplinary Approach.* New York: Springer Verlag, 1981, pp. 1-43.
2. Day, W. *Genesis on Planet Earth,* 2nd ed. New Haven: Yale University Press, 1984.

3. Williams, R. J. P. *Chemica Scripta* **1986**, *26*, 513.
4. Metzner, H. (ed.). *Photosynthetic Oxygen Evolution*. New York: Academic Press, 1978.
5. Ono, T. A.; Kajikawa, H.; Inoue, Y. *Plant Physiol* **1986**, *80*, 85.
6. Hoff, A. J. In *Light Reaction Path of Photosynthesis* (Fong, F. K., ed.). Heidelberg: Springer-Verlag, 1982; pp. 80-151.
7. Critchley, C.; Sargeson, A. M. *FEBS Lett.* **1984**, *177*, 2.
8. Dismukes, G. C.; Ambramowicz, D. A.; Ferris, K. F.; Mathur, P.; Siderer, Y.;
9. Hayaishi, D. (ed.). *Molecular Mechanisms of Oxygen Activation*. New York: Academic Press, 1974.
10. Fridovich, I. *Photochem. Photobiol.* **1979**, *28*, 733.
11. Wilshire, J.; Sawyer, D. T. *Acc. Chem. Res.* **1979**, *12*, 105.
12. Sawyer, D. T.; Valentine, J. S. *Acc. Chem. Res.* **1981**, *14*, 393.
13. Sugimoto, H.; Sawyer, D. T. *J. Am. Chem. Soc.* **1985**, *107*, 5712.
14. Sugimoto, H.; Sawyer, D. T. *J. Org. Chem.* **1985**, *50*, 1784.
15. Sawyer, D. T.; Roberts, J. L., Jr.; Calderwood, T. S.; Sugimoto, H.; McDowell, M. S. *Phil. Trans. Roy. Soc. Lond.* **1985**, *B311*, 483.

# 2

# REDOX THERMODYNAMICS FOR OXYGEN SPECIES (O₃, O₂, HOO·, O₂⁻·, HOOH, HOO⁻, O, O⁻·, AND HO⁻); EFFECTS OF MEDIA AND pH

*One electron at a time*

Biological systems activate ground-state dioxygen ($^3O_2$) for controlled energy transduction and chemical syntheses via electron-transfer and hydrogen-atom-transfer reduction to $O_2^{-\cdot}$, HOO·, and HOOH. These reduction products are further activated with metalloproteins to accomplish oxygen atom-transfer chemistry. Conversely, green plants via photosystem II facilitate the oxidation of water to $^3O_2$ (Chapter 1),[1]

$$2\,H_2O \xrightarrow[\text{PS-II [Mn, Ca, Cl}^-\text{, HOC(O)O}^-\text{]}]{4\,h\nu} {}^3O_2 + (4\,H^+ + 4\,e^-) \qquad (2\text{-}1)$$

which is responsible for the Earth's aerobic atmosphere (21% $O_2$). A small fraction of the $O_2$ in the upper atmosphere is transformed via solar radiation to ozone,

$$3\,O_2 \xrightarrow{h\nu} 2\,O_3 \qquad (2\text{-}2)$$

which provides a protective shield against short wave-length uv radiation.

## Redox thermodynamics for dioxgyen species and ozone

The reaction thermodynamics for these processes are influenced by the solution matrix and its acidity. Thus, the redox thermodynamics of $O_2$ are directly dependent upon proton activity,

$$O_2 + 4\,H_3O^+ + 4\,e^- \longrightarrow 2\,H_2O, \; E^{o\prime} \qquad (2\text{-}3)$$

which in turn depends upon the reaction matrix. Table 2-1 summarizes the $pK_a'$ values for a series of Brønsted acids in several aprotic solvents and water.[2] In acetonitrile the activity values for $pK_a'$ range from -8.8 for $(H_3O)ClO_4$ to 30.4 for $H_2O$. This means that the formal potential ($E^{o\prime}$) for reaction 2-3 in acetonitrile (MeCN) is +1.79 V versus NHE in the presence of 1 $M$ $(H_3O)ClO_4$ and -0.53 V in the presence of 1 $M$ $(Bu_4N)OH$.

Another limiting factor with respect to chemical-energy flux for oxidative metabolism and respiration is the solubility of $O_2$. Because of nonpoplar

Table 2-1   Effective $pK_a'$ Values for Brønsted Acids in Aprotic Solvents and Water[a]

| Brønsted acid | Solvent [0.5 M (Et$_4$N)ClO$_4$][b] | | | | |
| --- | --- | --- | --- | --- | --- |
|  | MeCN | DMF | Me$_2$SO | py | H$_2$O (lit) |
| (H$_3$O)ClO$_4$ | -8.8 | 0.7 | 2.6 | 4.6 | 0.0 (0.0) |
| $p$-MePhS(O)$_2$OH | -3.8 | – | – | – | – |
| (pyH)ClO$_4$ | 1.8 | 3.1 | 4.4 | 5.7 | 4.9 (5.2) |
| 2,4-(NO$_2$)$_2$PhOH | 4.3 | 4.5 | 4.9 | 5.5 | – (4.0) |
| (NH$_4$)ClO$_4$ | 5.9 | 9.6 | 11.7 | 7.3 | 8.7 (9.2) |
| (Et$_3$NH)Cl | 10.0 | 9.9 | 12.7 | 7.6 | 10.1 (11.0) |
| PhC(O)OH | 7.9 | 11.5 | 13.6 | 11.6 | 3.2 (4.2) |
| 2-Py-C(O)OH | 8.6 | – | – | – | – (5.5) |
| PhOH | 16.0 | 19.4 | 20.8 | 20.1 | 9.2 (9.9) |
| $p$-EtOPhOH | 19.3 | 21.5 | 23.8 | 21.8 | 9.6 |
| H$_2$O | 30.4 | 34.7 | 36.7 | 30.5 | 14.8 (14.0) |

[a] Ref. 2.
[b] MeCN, acetonitrile; DMF, dimethylformamide; Me$_2$SO, dimethyl sulfoxide; py, pyridine.

character dioxygen is much more soluble in organic solvents than in H$_2$O (Table 2-2).[3]

The reduction potentials for O$_2$ and various intermediate species in H$_2$O at $pH$ 0, 7, are 14 are summarized in Fig. 2-1[4-7]; similar data for O$_2$ in MeCN at $pH$ -8.8, 10.0, and 30.4 are presented in Fig. 2-2.[3,8,9] Potentials for the reduction of ozone (O$_3$) in aqueous and acetonitrile solutions are presented in Figure 2-3.[7,8]

The reduction manifolds for O$_2$ (Fig. 2-1 and 2-2) indicate that the limiting step (in terms of reduction potential) is the first electron transfer to O$_2$, and that an electron source adequate for the reduction of O$_2$ will produce all of the other reduced forms of dioxygen (O$_2^-$·, HOO·, HOOH, HOO$^-$, ·OH) via reduction, hydrolysis, and disproportionation steps (Scheme 2-1).[10,11] Thus, the most direct means to activate O$_2$ is the addition of an electron (or hydrogen atom), which results in significant fluxes of several reactive oxygen species.

The dominant characteristic of O$_2^-$· in any medium is its ability to act as a strong Brønsted base via formation of HOO·,[12,13] which reacts with itself or a second O$_2^-$· (Scheme 2-1). Within water, superoxide ion is rapidly converted to dioxygen and peroxide

$$2\ O_2^-\cdot + H_2O \rightleftharpoons O_2 + HOO^- + HO^-, \quad K \sim 2.5 \times 10^8\ \text{atm} \qquad (2\text{-}4)$$

Such a proton-driven disproportionation process means that O$_2^-$· can deprotonate acids much weaker than water (up to $pK_a \approx 23$).[14]

Table 2-2  Solubilities of $O_2$ (1 atm) in Various Solvents[a]

| Solvent | $[O_2]$ 1 atm (m$M$) |
|---|---|
| $H_2O$ | 1.0 |
| $Me_2SO$ | 2.1 |
| DMF | 4.8 |
| py | 4.9 |
| MeCN | 8.1 |
| Hydrocarbons | ~10 |
| Fluorocarbons | ~25 |

[a] Ref. 3.

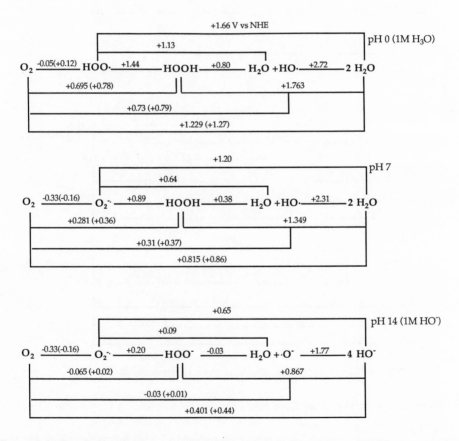

Figure 2-1  Standard reduction potentials for dioxygen species in water [$O_2$ at 1 atm] (formal potentials for $O_2$ at unit activity).

*Scheme 2-1   Formation and degradation of superoxide ion ($O_2^{-\cdot}$)*

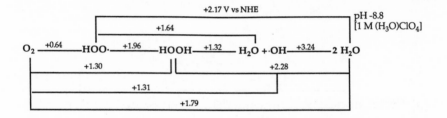

$$O_2 + e^- \xrightarrow{-0.33\ V} O_2^{-\cdot} \xrightarrow{HA} HOO\cdot + A^-$$

$$\xrightarrow{HOO\cdot} HOOH + O_2,\ k \sim 10^6\,M^{-1}s^{-1}$$

$$\xrightarrow{HA} HOOH + A^-$$

$$\xrightarrow{O_2^{-\cdot}} HOO^- + O_2,\ k \sim 10^8\,M^{-1}s^{-1}$$

$$\xrightarrow{HOOH} O_2^{-\cdot} + H_2O + HO\cdot$$

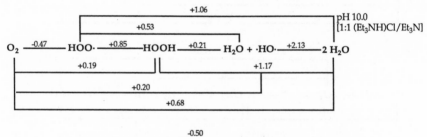

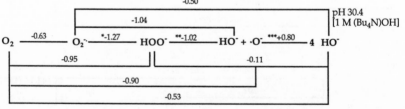

*(O$_2^{-\cdot}$ $\xrightarrow{-1.51}$ HOOH) **(HOOH $\xrightarrow{-0.90}$ HO$^-$ + HO·) ***(HO· $\xrightarrow{+0.92}$ HO$^-$)

Figure 2-2   Formal reduction potentials for dioxygen species in acetonitrile ($O_2$ at unit activity).

A. $H_2O$

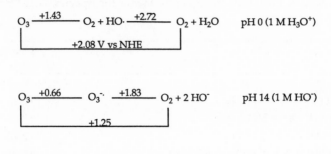

B. MeCN

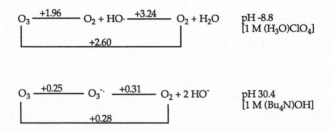

Figure 2-3 Standard reduction potentials for ozone in water and in acetonitrile.

The data of Figs. 2-1 and 2-2 indicate that $O_2^-$ is a moderate one-electron reducing agent [cytochrome-$c$(Fe$^{III}$) is reduced in $H_2O$[15,16] and iron(III) porphyrins in dimethylformamide].

$$(DMF)^+Fe^{III}TPP + O_2^- \longrightarrow Fe^{II}TPP + O_2, \ \Delta E, +0.7 \ V \qquad (2-5)$$

Superoxide ion reacts with proton sources to form HOO·, which disproportionates via a second $O_2^-$ or itself (Scheme 2-1). However, with limiting fluxes of protons to control the rate of HOO· formation from $O_2^-$, the rate of decay of HOO· is enhanced by reaction with the allylic hydrogens of excess 1,4-cyclohexadiene (1,4-CHD).[17] Because HOO· disproportionation is a second-order process, low concentrations favor hydrogen-atom abstraction from 1,4-CHD. This is especially so for Me$_2$SO, in which the rate of disproportionation for HOO· is the slowest (PhCl>MeCN>$H_2O$>DMF>Me$_2$SO).[17]

## Redox thermodynamics of atomic oxygen

The electron-transfer potentials for the reduction of ground-state atomic oxygen in water and in acetonitrile are summarized in Tables 2-3 and 2-4.[4,5,18] In addition, selected values for the reduction of the atomic oxygen in $O_3$, HOIO$_3$, HOOH, and HOCl are included.[4] Because the oxygen atom in a water molecule is

Table 2-3  Redox Potentials in Aqueous Media for Species That Contain Atomic Oxygen

|  | $E°$, V versus NHE |
|---|---|
| *p*H 0 |  |
| $O(g) + 2 H_3O^+ + 2 e^- \rightarrow 3 H_2O$ | +2.43 |
| $O(g) + H_3O^+ + e^- \rightarrow HO\cdot + H_2O$ | +2.14 |
| $HO\cdot + H_3O^+ + e^- \rightarrow 2 H_2O$ | +2.72 |
| $O_3(g) + 2 H_3O^+ + 2 e^- \rightarrow O_2(g) + 3 H_2O$ | +2.08 |
| $O_3(g) + H_3O^+ + e^- \rightarrow HO\cdot + O_2 + H_2O$ | +1.43 |
| $HOOH + 2 H_3O^+ + 2 e^- \rightarrow 4 H_2O$ | +1.76 |
| $HOIO_3 + H_3O^+ + 2 e^- \rightarrow IO_3^- + 2 H_2O$ | +1.6 |
| $HOCl + 2 H_3O^+ + 2 e^- \rightarrow HCl + 3 H_2O$ | +1.49 |
| *p*H 7 |  |
| $O(g) + 2 H_3O^+ + 2 e^- \rightarrow 3 H_2O$ | +2.01 |
| $O(g) + H_3O^+ + e^- \rightarrow HO\cdot + H_2O$ | +1.71 |
| $HO\cdot + H_3O^+ + e^- \rightarrow 2 H_2O$ | +2.31 |
| $O_3(g) + H_3O^+ + 2 e^- \rightarrow O_2(g) + 3 H_2O$ | +1.66 |
| $HOOH + H_3O^+ + 2 e^- \rightarrow 4 H_2O$ | +1.35 |
| $IO_4^- + 2 H_3O^+ + 2 e^- \rightarrow IO_3^- + 3 H_2O$ | +1.2 |
| *p*H 14 |  |
| $O(g) + H_2O + 2 e^- \rightarrow 2 HO^-$ | +1.60 |
| $O(g) + H_2O + e^- \rightarrow HO\cdot + HO^-$ | +1.31 |
| $O(g) + e^- \rightarrow O^{-\cdot}$ | +1.43 |
| $HO\cdot + e^- \rightarrow HO^-$ | +1.89 |
| $HOH + e^- \rightarrow H\cdot + HO^-$ | -2.93 |
| $O^{-\cdot} + H_2O + e^- \rightarrow 2 HO^-$ | +1.77 |
| $O_3(g) + H_2O + 2 e^- \rightarrow O_2(g) + 2 HO^-$ | +1.25 |
| $O_3(g) + e^- \rightarrow O_3^{-\cdot}$ | +0.66 |
| $O_3^{-\cdot} + H_2O + e^- \rightarrow O_2(g) + 2 HO^-$ | +1.83 |
| $HOO^- + H_2O + 2 e^- \rightarrow 3 HO^-$ | +0.87 |
| $IO_4^- + H_2O + 2 e^- \rightarrow IO_3^- + 2 HO^-$ | +0.8 |
| $ClO^- + H_2O + 2 e^- \rightarrow Cl^- + 2 HO^-$ | +0.89 |
| $ClO^- + e^- \rightarrow Cl^- + O^{-\cdot}$ | +0.02 |

essentially without charge, these reduction processes represent the facilitated reduction of protons via stabilization of the product H atom through strong covalent-bond formation with atomic oxygen.    Thus,  the  standard-state reduction of protons to hydrogen atoms is

$$2 H_3O^+ + 2 e^- \longrightarrow 2 H\cdot + 2 H_2O, \quad E°, \text{-}2.10 \text{ V versus NHE} \qquad (2\text{-}6)$$

Table 2-4  Redox Potentials for Species in MeCN [$H^+ \equiv 1\ M$ ($H_3O$)$ClO_4$; $HO^- \equiv 1\ M$ ($Bu_4N$)$OH$($MeOH$)] That Contain Atomic Oxygen

| | $E°$, V versus NHE |
|---|---|
| Acid, $pH$ (-8.8): | |
| $O(g) + 2\ H^+ + 2\ e^- \rightarrow H_2O$ | +2.95 |
| $O(g) + H^+ + e^- \rightarrow HO\cdot$ | +2.66 |
| $HO\cdot + H^+ + e^- \rightarrow H_2O$ | +3.24 |
| $O_3(g) + 2\ H^+ + 2\ e^- \rightarrow O_2(g) + H_2O$ | +2.60 |
| $HOOH + 2\ H^+ + 2\ e^- \rightarrow 2\ H_2O$ | +2.28 |
| $HOCl + 2\ H^+ + 2\ e^- \rightarrow HCl + H_2O$ | +2.0 |
| Base, $pH$ 30.4 | |
| $O(g) + H_2O + 2\ e^- \rightarrow 2\ HO^-$ | +0.63 |
| $O(g) + H_2O + e^- \rightarrow HO\cdot + HO^- \rightarrow O^-\cdot(H_2O)$ | +0.34 |
| $HO\cdot + e^- \rightarrow HO^-$ | +0.92 |
| $O^-\cdot + H_2O + e^- \rightarrow 2\ HO^-$ | +0.80 |
| $HOH + e^- \rightarrow H\cdot + HO^-$ | -3.90 |
| $O_3 + H_2O + 2\ e^- \rightarrow O_2 + 2\ HO^-$ | +0.28 |
| $ClO^- + H_2O + 2\ e^- \rightarrow Cl^- + 2\ HO^-$ | -0.08 |
| $HOO^- + H_2O + 2\ e^- \rightarrow 3\ HO^-$ | -0.10 |

shifted to a much more favored process in the presence of atomic oxygen

$$O(g) + 2\ H_3O^+ + 2\ e^- \longrightarrow 3\ H_2O,\ E°,\ +2.43\ \text{V versus NHE} \qquad (2\text{-}7)$$

Likewise, the electron-transfer reduction of $H_2O$ (with uncharged hydrogen and oxygen atoms), which must overcome the stabilization of the strong O–H bonds, results in a -1 charge for oxygen rather than proton reduction

$$H_2O + e^- \longrightarrow H\cdot + HO^-,\ E°,\ +2.93\ \text{V versus NHE} \qquad (2\text{-}8)$$

In contrast, an equivalent charge-density change for the neutral oxygen in $HO\cdot$ is strongly favored[9]

$$HO\cdot + e^- \longrightarrow HO^-,\ E°,\ +1.89\ \text{V versus NHE} \qquad (2\text{-}9)$$

and reduction of the combination of the zero-charge oxygen of water with -1-charge $O^-\cdot$ yields two -1-charge hydroxide ions[5]

$$O^-\cdot + H_2O + e^- \longrightarrow 2\ HO^-,\ E°,\ +1.77\ \text{V} \qquad (2\text{-}10)$$

Although the chlorine atom is among the strongest one-electron oxidants known,[5]

$$\text{Cl} \cdot + e^- \longrightarrow \text{Cl}^-, \ E^\circ, \ +2.41 \text{ V versus NHE} \tag{2-11}$$

when it is stabilized by the covalent bond of the $\text{Cl}_2$ molecule or of the $\text{ClO}^-$ ion its propensity to add electrons is reduced[4,5]

$$\text{Cl}_2 + e^- \longrightarrow \text{Cl}_2^{-\cdot}, \ E^\circ, \ +0.63 \text{ V} \tag{2-12}$$

$$\text{Cl}_2 + 2 \ e^- \longrightarrow 2 \ \text{Cl}^-, \ E^\circ, \ +1.36 \text{ V} \tag{2-13}$$

$$\text{ClO}^- + e^- \longrightarrow \text{Cl}^- + \text{O}^{-\cdot}, \ E^\circ, \ +0.20 \text{ V} \tag{2-14}$$

$$\text{ClO}^- + \text{H}_2\text{O} + 2 \ e^- \longrightarrow \text{Cl}^- + 2 \ \text{HO}^-, \ E^\circ, \ +0.89 \text{ V} \tag{2-15}$$

Thus, $\text{Cl}_2$ can transfer an electron from an $\text{HO}^-$ because of the stabilization of the resulting $\text{HO}\cdot$ via bond formation with an $\text{H}\cdot$ from a second $\text{HO}^-$; the latter product ($\text{O}^{-\cdot}$) is stabilized via bond formation with a $\text{Cl}\cdot$ [$\Delta E^\circ$, +0.47 V (Eqs. 2-13 – 2-15)]

$$\text{Cl}_2 + 2 \ \text{HO}^- \longrightarrow \text{ClO}^- + \text{Cl}^- + \text{H}_2\text{O} \tag{2-16}$$

### Nucleophilic attack via single-electron transfer

The facile nucleophilic displacement of chloride from alkyl chlorine compounds (all uncharged atoms) by $\text{HO}^-$ (-1-charge O) to give an alcohol (all uncharged atoms) is a charge-transfer process via electron transfer from the $\text{HO}^-$ to the chlorine atom of $R\text{Cl}$[19,20]

$$n\text{-BuCl} + \text{HO}^- \longrightarrow n\text{-BuOH} + \text{Cl}^- \tag{2-17}$$

Likewise the nucleophilic addition of $\text{HO}^-$ (-1-charge O) to $\text{CO}_2$ gives bicarbonate ion with the charge of the oxygen of the C–OH group changed to zero via electron transfer (and charge transfer) to a neutral carbonyl oxygen

$$\text{O=C=O} + \text{HO}^- \longrightarrow \overset{\displaystyle \text{O}}{\overset{\displaystyle \|}{\text{HO-C-O}^-}} \tag{2-18}$$

The data of Figs. 2-1 and 2-2 and of Tables 2-3 and 2-4 provide persuasive evidence that the electron-transfer and atom-transfer chemistry for dioxygen species and mono-oxygen species is strongly dependent upon proton activity, anion solvation, and relative X–OO and Y–O covalent bond energies. Hence, the solvation energy of $\text{HO}^-$ is 22.5 kcal less in acetonitrile than in water, which causes the $\text{HO}\cdot/\text{HO}^-$ redox potential to be +0.92 V versus NHE in MeCN instead of +1.89 V versus NHE in $\text{H}_2\text{O}$.

## Electrochemistry of dioxygen

The ground state of molecular oxygen ($\cdot O_2\cdot$, $^3\Sigma_g^-$) has two unpaired electrons in degenerate $2\pi_g$ orbitals (Table 3-3, Chapter 3), and is referred to as *dioxygen* by most contemporary biologists and biochemists. When dioxygen is reduced by electron transfer, a series of intermediate basic dioxygen and mono-oxygen species are produced that may take up one or two protons from the media ($O_2^-\cdot$, $HOO\cdot$, $HOO^-$, $HOOH$, $\cdot O^-$, $HO\cdot$, $HO^-$, and $H_2O$). The thermodynamics for the various reduction steps in aqueous solutions have been evaluated by numerous techniques, but all are fundamentally based on the calorimetry associated with the reaction

$$O_2(g) + 2 H_2(g) \longrightarrow 2 H_2O(l), \quad -\Delta G^\circ = nE^\circ_{cell}F \tag{2-19}$$

If redox potentials are relative to the normal hydrogen electrode,

$$2 H_3O^+ + 2 e^- \longrightarrow H_2(g), \quad E^\circ_{H^+/H_2} \equiv 0.000 \tag{2-20}$$

then the standard redox potential for the four-electron reduction of dioxygen ($E^\circ_{O_2/H_2O}$) can be calculated from the calorimetric data for the reaction of Eq. (2-19) under standard conditions

$$O_2(g) + 4 H^+(aq) + 4 e^- \longrightarrow 2 H_2O(l), \quad E^\circ_{O_2/H_2O} \equiv$$
$$E^\circ_{cell} - E^\circ_{H^+/H_2} = +1.229 \text{ V versus NHE} \tag{2-21}$$

Figure 2-1 summarizes the redox potentials for the reduction of various dioxygen species in aqueous media at $pH$ 0, 7, and 14. For those couples that involve dioxygen itself, formal potentials are given in parentheses for $O_2$ at unit activity (~$10^3$ atm; $[O_2] \approx 1$ m$M$ at 1 atm partial pressure).

Although an early voltammetric investigation demonstrated that dioxygen in aqueous solutions is reduced to hydrogen peroxide at platinum electrodes,[21]

$$O_2 + 2 H_3O^+ + 2 e^- \longrightarrow HOOH + 2 H_2O, \quad E_{p,c}, +0.05 \text{ V versus NHE} \tag{2-22}$$

the $pH$ independence of the reduction potential and the irreversibility of the process confirm that this represents the overall reduction process and that the mechanism is *not* a single, concerted, two-electron reduction. Subsequent voltammetric studies have established that the reduction process depends upon the surface of solid electrodes in aqueous media.[22,23]

The $pH$-dependent reduction potential for $O_2$ that is observed with activated metal surfaces is due to the reduction of the freshly formed metal hydroxide film; for example,

$$Pt(OH)_2 + 2 H_3O^+ + 2 e^- \longrightarrow Pt + 4 H_2O, \quad E^{\circ\prime}, +0.67 \text{ V versus NHE} \tag{2-23}$$

After the metal oxide is reduced electrochemically, it is reformed by chemical reaction with $O_2$

$$2\,Pt + O_2 + 2\,H_2O \longrightarrow 2\,Pt(OH)_2 \qquad (2\text{-}24)$$

Hence, the metal oxide is a catalyst for the reduction of $O_2$ to $H_2O$. Such a mechanism accounts for the dependence of the reduction potentials and the rates of reaction on the electrode material and $p$H.

The older literature on the electrochemistry of dioxygen in acidic media attributes the difference between the thermodynamic potential for the four-electron reduction ($O_2/H_2O$ , +1.23 V versus NHE; Fig. 2-1) and the observed value at a freshly activated platinum electrode [+0.67 V versus NHE, Eq. (2-23)] to overvoltage (or kinetic inhibition).  Likewise, the difference between the thermodynamic potential for the two-electron reduction ($O_2/HOOH$, +0.70 V versus NHE, Fig. 2-1) and the observed value at passivated electrodes [+0.05 V versus NHE, Eq. (2-22)] was believed to be due to the kinetic inhibition of the two-electron process.

*Electron-transfer reduction of $O_2$.*

Within aqueous solutions the most direct means to the electron-transfer reduction of dioxygen is by pulse radiolysis.  Irradiation of an aqueous solution by an electron beam yields (almost instantly) solvated electrons [$e^-$(aq)], hydrogen atoms (H·), and hydroxyl radicals (HO·).  If the solution contains a large excess of sodium formate $Na^+ \cdot O(O)CH$ and is saturated with $O_2$, then the radiolytic electron flux efficiently and cleanly reduces $O_2$ to superoxide ion ($O_2^{-\cdot}$).[24–28]

$$\cdot O_2 \cdot + e^-(aq) \longrightarrow O_2^{-\cdot} \qquad (2\text{-}25)$$

$$HC(O)O^- + HO\cdot \longrightarrow \cdot C(O)O^- + H_2O \qquad (2\text{-}26)$$

$$\cdot O_2 \cdot + \cdot C(O)O^- \longrightarrow O_2^{-\cdot} + CO_2 \qquad (2\text{-}27)$$

$$\cdot O_2 \cdot + H\cdot \longrightarrow HOO\cdot \xrightarrow{HO^-} H_2O + O_2^{-\cdot} \qquad (2\text{-}28)$$

This represents electrodeless electrochemistry, whereby the only process is electron transfer and the only product is the one-electron adduct of $O_2$, the superoxide ion ($O_2^{-\cdot}$).  Through the use of redox mediator dyes and spectrophotometry, the electron-transfer thermodynamic reduction potential for $O_2$ has been evaluated

$$O_2(g,\,1\ atm) + e^- \rightleftharpoons O_2^{-\cdot},\ (E^{\circ\prime})_{pH\ 5\text{-}14}\ ,\ -0.33\ V\ \text{versus NHE} \qquad (2\text{-}29)$$
$$-0.16\ V\ \text{versus NHE for}$$
$$O_2\ \text{at unit activity}$$

This relation in combination with the $O_2/HOOH$ couple at $pH$ 7.0 of Fig. 2-1 yields the one-electron reduction potential for $O_2^{-\cdot}$.

$$O_2^{-\cdot} + H_3O^+ + e^- \rightleftharpoons HOOH, \quad (E^{o\prime})_{pH\,7}, +0.89 \text{ V versus NHE} \qquad (2\text{-}30)$$

A direct electrochemical measurement of the reduction potential for the $O_2/O_2^{-\cdot}$ couple in aqueous solutions is complicated by the rapid proton-induced disproportionation reactions for $O_2^{-\cdot}$. The perhydroxyl radical (HOO·) is a moderately weak acid (roughly equivalent to acetic acid)[24]

$$HOO\cdot + H_2O \rightleftharpoons H_3O^+ + O_2^{-\cdot}, \quad pK_a, 4.89 \qquad (2\text{-}31)$$

which undergoes a rapid homolytic disproportionation

$$HOO\cdot + HOO\cdot \xrightarrow{\;k_{bi},\, 8.6 \times 10^5 \; M^{-1}s^{-1}\;} HOOH + O_2, \quad K, 10^{49} \qquad (2\text{-}32)$$

as well as an even faster process with $O_2^{-\cdot}$.[25–28]

$$HOO\cdot + O_2^{-\cdot} \xrightarrow[\;H_3O^+\;]{\;k_{bi},\, 1.0 \times 10^8 \; M^{-1}s^{-1}\;} HOOH + O_2 + H_2O \qquad (2\text{-}33)$$

Hence, the maximum rate of disproportionation occurs at $pH$ 4.9 (the $pK_a$ for HOO·).

Figure 2-4 illustrates the electrochemical reduction of $O_2$ at platinum electrodes in aqueous media (1.0 $M$ $NaClO_4$). The top curve represents the cyclic voltammogram (0.1 V s$^{-1}$) for $O_2$ at 1 atm (~1 m$M$), and the lower curve is the voltammogram with a rotated-disk electrode (900 rpm, 0.5 V min$^{-1}$). Both processes are totally irreversible with two-electron stoichiometries and half-wave potentials ($E_{1/2}$) that are independent of $pH$. The mean of the $E_{1/2}$ values for the forward and reverse scans of the rotated-disk voltammograms for $O_2$ is 0.0 V versus NHE. If the experiment is repeated in media at $pH$ 12, the mean $E_{1/2}$ value also occurs at 0.0 V.

The aqueous electrochemical reduction of $O_2$ at inert electrodes occurs via an initial reversible electron-transfer process [analogous to the homogeneous process of pulse radiolysis, Eq. (2-29)]

$$O_2 + e^- \rightleftharpoons O_2^{-\cdot}, \quad E_{1/2}\,(pH\,12)\,, 0.0 \text{ V versus NHE} \qquad (2\text{-}34)$$

which is followed by two rapid chemical steps

$$O_2^{-\cdot} + H_2O \rightleftharpoons HOO\cdot + HO^-, \quad K = 10^{-14}/10^{-4.9} = 8 \times 10^{-10} \qquad (2\text{-}35)$$

$$HOO\cdot + HOO\cdot \rightleftharpoons HOOH + O_2, \quad K, 10^{49} \qquad (2\text{-}36)$$

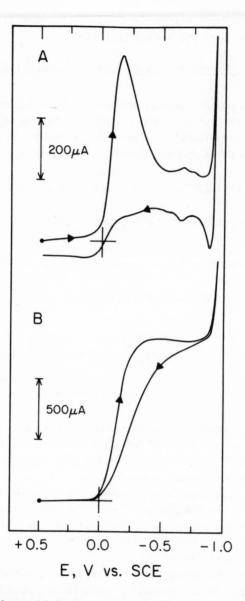

Figure 2-4 Electrochemical behavior of dioxygen (1 atm) in aqueous solutions (1 *M* NaClO$_4$) at a platinum electrode (area, 0.458 cm$^2$). (a) The cyclic voltammogram was initiated at the rest potential with a scan rate of 0.1 V s$^{-1}$. (b) The rotated-disk (400 rpm) voltammogram was obtained with a scan rate of 0.5 V min$^{-1}$.

The rate and extent of the hydrolysis reaction [Eq. (2-35)] is suppressed under alkaline conditions, and the lifetime of O$_2^{-\cdot}$ is increased such that the $E_{1/2}$ potential may be shifted to more negative potentials. This is the result of thermodynamic effects via the Nernst equation rather than kinetic (overvoltage) effects. The overall post electron-transfer chemistry is given by the relation

$$2\,O_2^{-\cdot} + 2\,H_2O \; \overrightarrow{\phantom{xxxxxxxx}}\!\!\!\!\!\overleftarrow{\phantom{xxxxxxxx}} \; HOOH + O_2 + 2\,HO^-, \; K, 6.4 \times 10^{30} \tag{2-37}$$

The net electrochemical process is the sum of Eqs. (2-34) and (2-37),

$$O_2 + 2\,H_2O + 2\,e^- \longrightarrow HOOH + 2\,HO^- \tag{2-38}$$

and is observed with inert electrodes in aqueous media. With metal electrodes there is a tendency for the intermediate species ($O_2^{-\cdot}$ and $HOO\cdot$ ) to react and form electroactive metal oxides.

*Aprotic media*

In the absence of proton sources dioxygen is reversibly reduced to superoxide ion, as illustrated by Fig. 2-5 for a pyridine solution (0.1 $M$ TEAP).[29,30]

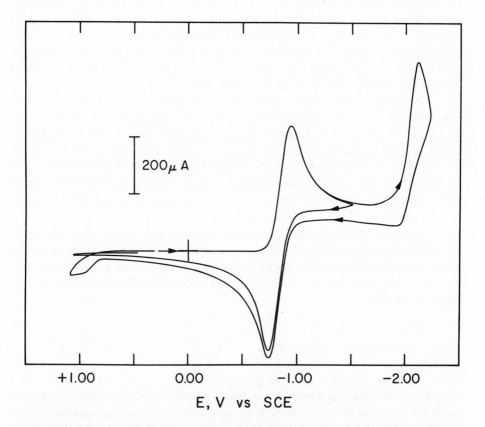

Figure 2-5 Cyclic voltammogram for 5 m$M$ dioxygen ($O_2$, 1 atm) in pyridine (0.1 M tetrapropylammonium perchlorate) at a platinum electrode (area, 0.23 cm$^2$). Scan rate, 0.1 V s$^{-1}$. Saturated calomel electrode (SCE) versus NHE, +0.242 V.

$$O_2 + e^- \rightleftharpoons O_2^{-}\cdot, \quad E^{\circ\prime}, \text{ -0.64 V versus NHE} \tag{2-39}$$

The second reduction is an irreversible one-electron process

$$O_2^{-}\cdot + e^- \xrightarrow{\ H_2O\ } HOO^- + HO^-, \quad E_{p,c}, \text{ -1.8 V versus NHE} \tag{2-40}$$

$$HO^- + HOO^- + O_2 \rightleftharpoons 2\,O_2^{-}\cdot + H_2O \tag{2-41}$$

In alkaline pyridine solutions HOOH also decomposes to give $O_2^{-}\cdot$.[31]

$$HOO^- + HOOH \xrightarrow[k, 10^4\ M^{-1}s^{-1}]{\ py\ } O_2^{-}\cdot + H_2O + \frac{1}{n}[py(OH)]_n \tag{2-42}$$

Although there is a general view that the electrochemical reduction of $O_2$ in aprotic media is independent of media and electrode materials, the cyclic voltammograms of Fig. 2-6 and the electrochemical data of Table 2-5 provide clear evidence that both have a significant effect on the reversibility. Although the peak separation ($\Delta E_p$) varies with solvent and with electrode material, the median potentials $(E_{p,c} + E_{p,a})/2$ are essentially independent of electrode material and provide a reasonable measure for the formal reduction potential ($E^{\circ\prime}$) for the $O_2/O_2^{-}\cdot$ couple. The average values for $E^{\circ\prime}$ are summarized in Table 2-5 together with the value for an aqueous solvent at $pH$ 7.

Reference to Table 2-5 confirms that the reduction potential for the $O_2/O_2^{-}\cdot$ couple shifts to more negative values as the solvating properties of the media decrease. The heat of hydration $(-\Delta H_{aq})$ for gaseous $O_2^{-}\cdot$ is 418 kJ (100 kcal), which is consistent with the unique strong solvation of anions by water. Hence, if the $E^{\circ\prime}_{O_2/O_2^{-}\cdot}$ values for the $O_2/O_2^{-}\cdot$ couple are affected primarily by the degree of solvation of $O_2^{-}\cdot$ (that is, the solvation energy for $O_2$ is assumed to be small and about the same for different solvents), then the relative solvation energies for $O_2^{-}\cdot$ are $H_2O \gg Me_2SO > DMF > py \sim MeCN$.

The variation in the peak-separation values ($\Delta E_p$) for the cyclic voltammetric data of Table 2-5 may be interpreted in terms of heterogeneous electron-transfer kinetics, but the most reasonable explanation is uncompensated resistance (especially for py and MeCN) and surface reactions (especially for the metal electrodes).

*Protic and electrophilic substrates*

When dioxygen is reduced in the presence of an equimolar concentration of a strong acid ($HClO_4$) in dimethylformamide (DMF) (Fig. 2-7), a new irreversible peak occurs at -0.13 V versus SCE in addition to the regular quasireversible couple at -0.86 V.[3] [In the absence of $O_2$, strong acids ($HClO_4$) in DMF exhibit a reversible one-electron couple at -0.37 V.] The peak height at -0.13 V increases linearly as the concentration of $HClO_4$ is increased up to a mole ratio of 4:1

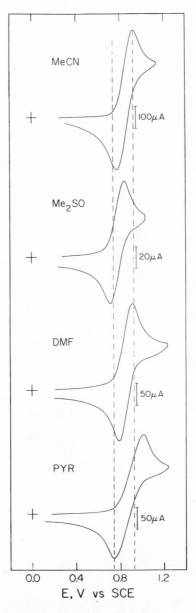

Figure 2-6  Cyclic voltammograms for $O_2$ (at 1 atm) in acetonitrile (MeCN) (8.1 m$M$), dimethyl sulfoxide (Me$_2$SO) (2.1 m$M$), dimethylformamide (DMF) (4.8 m$M$), and pyridine (py) (4.9 m$M$) at a vitreous carbon inlay electrode (area, 0.20 cm$^2$).  All of the solutions contained 0.1 $M$ TEAP as supporting electrolyte, and the scan rate was 0.1 V s$^{-1}$.

relative to $O_2$ (equivalent to a two-electron reduction).  This increase in peak height is at the expense of the reversible couple at -0.86 V.  The addition of excess phenol to a dioxygen/DMF solution causes the one-electron process to become a two-electron reduction to yield HOOH as the major product (Fig. 2-7); the reverse

Table 2-5 Redox Potentials for the Reduction of Dioxygen (1 atm) by Cyclic Voltammetry with Four Different Electrodes in Four Aprotic Solvents (0.1 $M$ Tetraethylammonium Perchlorate)

a. Electrochemical peak potentials; scan rate, 0.1 V s$^{-1}$; V versus NHE

| Solvent | C | | | Pt | | | Au | | | Hg | | |
|---|---|---|---|---|---|---|---|---|---|---|---|---|
| | $E_{p,c}$ | $E_{p,a}$ | $\Delta E_p$ | $E_{p,c}$ | $E_{p,a}$ | $\Delta E_p$ | $E_{p,c}$ | $E_{p,a}$ | $\Delta E_p$ | $E_{p,c}$ | $E_{p,a}$ | $\Delta E_p$ |
| Me$_2$SO | -0.62 | -0.46 | 0.16 | -0.87 | -0.70 | 0.17 | -0.66 | -0.44 | 0.22 | | | |
| DMF | -0.71 | -0.53 | 0.18 | -0.72 | -0.52 | 0.20 | -0.74 | -0.54 | 0.20 | | | |
| Py | -0.76 | -0.52 | 0.24 | -0.79 | -0.52 | 0.27 | -0.76 | -0.51 | 0.25 | | | |
| MeCN | -0.71 | -0.54 | 0.17 | -0.94 | -0.36 | 0.56 | -0.88 | -0.41 | 0.47 | -0.88 | -0.37 | 0.51 |

b. Formal reduction potentials for the O$_2$/O$_2$$^-$ couple (molar concentrations for O$_2$ and O$_2$$^-$)

| | E°, V versus NHE | | E°, V versus NHE |
|---|---|---|---|
| H$_2$O | -0.16 | py | -0.64 |
| Me$_2$SO | -0.54 | MeCN | -0.63 |
| DMF | -0.62 | | |

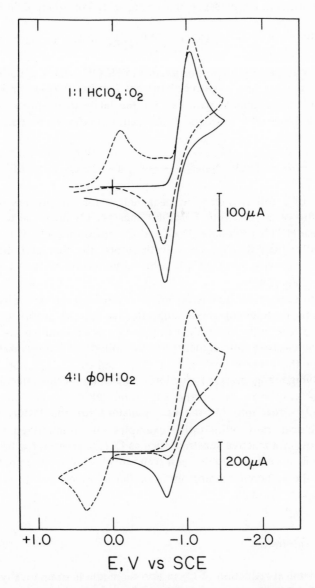

Figure 2-7 Cyclic voltammograms for 4.8 mM $O_2$ (at 1 atm) in DMF (0.1 $M$ TEAP) (solid lines) and in the presence of an equimolar concentration of $HClO_4$ and in the presence of a 4:1 mole ratio of phenol (PhOH) (dashed lines); platinum electrode, area 0.23 cm$^2$; scan rate, 0.1 V s$^{-1}$.

scan indicates that phenoxide is formed during the reduction of $O_2$. Similar effects are observed when other moderately weak acids are present in excess, but the addition of a fivefold excess (relative to $O_2$ concentration) of $H_2O$ or 1-butanol does not affect the electrochemistry of $O_2$ in DMF.

The cyclic voltammogram of Fig. 2-7 indicates that a new process occurs when $O_2$ is reduced in the presence of a strong acid. For example, in DMF

$$O_2 + (DMF)H^+ + e^- \longrightarrow HOO\cdot + DMF, \quad E_{p,c}, -0.11 \text{ V versus NHE} \qquad (2\text{-}43)$$

The HOO· species rapidly disproportionates to HOOH, which is not electroactive in DMF at potentials less negative than that for the $O_2/O_2^{-\cdot}$ couple (-0.64 V versus NHE). A reasonable pathway for the facile disproportionation is the formation of a dimer with subsequent dissociation to dioxygen and peroxide (see Chapter 5).[17]

$$HO_2\cdot + HO_2\cdot \longrightarrow [\text{cyclic dimer}] \longrightarrow HOOH + O_2, \quad k > 10^4 \, M^{-1} \, s^{-1} \qquad (2\text{-}44)$$

Achievement of a full two-electron peak height for the process of Eq. (2-43) requires a ratio of at least four $(DMF)H^+$ ions per $O_2$ molecule in DMF. This results because $HClO_4$ protonates DMF and the diffusion coefficient for the latter is much smaller than that for $O_2$. Furthermore, the flux of $(DMF)H^+$ to the electrode must be twice that for $O_2$ to achieve the second cycle of Eq. (2-43) with the products of Eq. (2-44).

Because superoxide ion is an effective nucleophile (see Chapter 7), the presence of electrophilic substrates with effective leaving groups (alkyl halides, acyl halides, esters, and anhydrides) in a dioxygen solution causes an apparent increase in the electron stoichiometry for the cathodic voltammogram of $O_2$ and a decrease in the reverse peak height (these substrates are not electroactive within the voltage range for $O_2$ reduction). The extent of this effect is dependent upon (1) the concentration of the substrate, (2) the stoichiometry for the substrate–$O_2^{-\cdot}$ reaction, (3) the rate constant for the latter, and (4) the voltammetric scan rate. Illustrative examples are summarized in Table 2-6. When an excess of a reactive substrate such as $CCl_4$ is present, the peak height for $O_2$ reduction is more than doubled at a scan rate of 0.1 V s$^{-1}$ (the process is totally irreversible); the cyclic voltammogram is similar in appearance to that for the 4:1 PhOH/$O_2$ system (Fig. 2-7).

*Metal cation substrates*

The electrochemical reduction of $O_2$ in aprotic media is dramatically changed by the presence of electroinactive metal cations.[3] Figure 2-8 illustrates the effect of a five-fold excess of $[Zn^{II}(H_2O)_6](ClO_4)_2$, $[Zn^{II}(\text{dimethylurea})_6](ClO_4)_2$, and $[Zn^{II}(\text{bipy})_3](ClO_4)_2$ on the cyclic voltammetry of $O_2$ in DMF at a platinum electrode. Prior to each reductive scan the electrode has been repolished; a second scan yields a much reduced peak current. In the presence of an excess concentration of Zn(II) cations [as well as the cations of Cd(II), Fe(II), Mn(II), and Co(II)], the reduction of $O_2$ is a totally irreversible process, and the electrodes (Pt, Au and C) are passivated after the initial negative scan.

For slow scan rates in the presence of 10- to 30-fold excess concentrations (relative to the $O_2$ concentration) of several divalent metal cations, the potentials

Table 2-6 Examples of Organic Substrates that Enhance the Voltammetric Peak Height for the Reduction of $O_2$ via Post-Electron-Transfer Reactions by $O_2^-$ in DMF (the Source of $O_2^-$ is the Reduction Process $O_2 + e^- \rightarrow O_2^-$ )

---

a. MeCl

    Rate-controlling step      $MeCl + O_2^- \rightarrow MeOO\cdot + Cl^-$, $k$, 80 $M^{-1}$ $s^{-1}$

    Overall stoichiometry      $2\,MeCl + 2\,O_2^- \rightarrow MeOOMe + 2\,Cl^- + O_2$

    Electron stoichiometry      $2\,MeCl + O_2 + 2\,e^- \rightarrow MeOOMe + 2\,Cl^-$;

    (in the limit as $v \rightarrow 0$)[a]      $2\,e^-/O_2$

b. $CCl_4$

    Rate-controlling step      $CCl_4 + O_2^- \rightarrow Cl_3COO\cdot + Cl^-$, $k$, 1300 $M^{-1}$ $s^{-1}$

    Overall stoichiometry      $CCl_4 + 6\,O_2^- \rightarrow CO_4^{2-} + 4\,Cl^- + 4\,O_2$

    Electron stoichiometry      $CCl_4 + 2\,O_2 + 6\,e^- \rightarrow CO_4^{2-} + 4\,Cl^-$; 3 $e^-/O_2$

    (in the limit as $v \rightarrow 0$)

c. MeC(O)OPh

    Rate-controlling step      $MeC(O)OPh + O_2^- \rightarrow MeC(O)OO\cdot + PhO^-$;

         $k$, 160 $M^{-1}$ $s^{-1}$

    Overall stoichiometry      $2\,MeC(O)OPh + 4\,O_2^- \rightarrow 2\,MeC(O)O^- +$

         $2\,PhO^- + 3\,O_2$

    Electron stoichiometry      $2\,MeC(O)OPh + O_2 + 4\,e^- \rightarrow 2\,MeC(O)O^- +$

    (in the limit as $v \rightarrow 0$)      $2\,PhO^-$; 4 $e^-/O_2$

---

[a] $v$ represents the voltammetric scan rate at a platinum electrode (0.23 $cm^2$) for the reduction of $O_2$ in the presence of excess organic substrate (Ref. 3).

for $O_2$ reduction are shifted to more positive values and the peak currents are increased by a factor of two or four. Table 2-7 summarizes the results for such experiments with platinum electrodes in $Me_2SO$ (0.1 M TEAP). Apparently, these cations act as Lewis acids relative to the $O_2^-$ product species from the electron-transfer step and thereby cause the reduction process to be shifted to more positive potentials in the order: $Fe^{2+} > Mn^{2+} > Co^{2+} \sim Zn^{2+} \sim Cd^{2+} \sim Li^+ \sim TEA^+$ in [tetraethylammonium cation from the supporting electrolyte, TEA($ClO_4$)]. On the basis of the electron stoichiometries of Table 2-7, the Lewis acid–superoxide adducts $[M^{2+}\text{-}(O_2^-)_n]$ disproportionate to metal peroxides and dioxygen. In the case of $Mn^{2+}$ and $Co^{2+}$ their peroxides are unstable and disproportionate to metal oxides and dioxygen to yield an overall stoichiometry of four electrons per $O_2$.

    The presence of metal cations also promotes post-electron-transfer chemistry with the $O_2^-$ from the $O_2$ reduction. The results of Fig. 2-8 and Table 2-7 can be rationalized on the basis of a Lewis acid–base process that promotes disproportionation to metal peroxide and $O_2$. For example, with excess zinc ion relative to $O_2$ in aprotic media

$$Zn^{II}(OH_2)_4^{2+} + O_2^- \longrightarrow (H_2O)_4^+ Zn^{II}(O_2) \qquad (2\text{-}45)$$

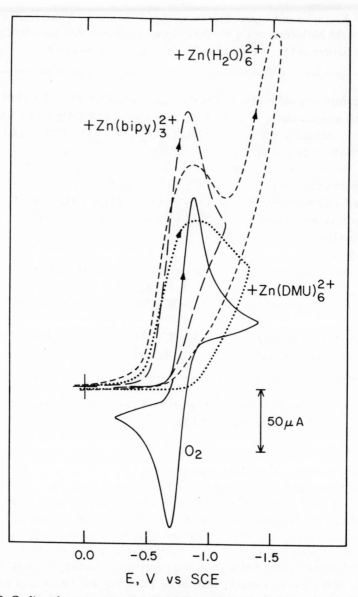

Figure 2-8  Cyclic voltammograms in Me₂SO (0.1 *M* TEAP) of 2.1 m*M* O₂ (at 1 atm) (solid line) and in the presence of 50 m*M* Zn$^{II}$(DMU)₆(ClO₄)₂ (DMU, dimethylurea) (dotted line), 50 m*M* Zn$^{II}$(H₂O)₆(ClO₄)₂ (bipy, 2,2'-bipyridine) (long dashed line); platinum electrode, area 0.23 cm²; scan rate 0.1 V s⁻¹.

$$(H_2O)_4^+Zn^{II}(O_2) + O_2^{-\cdot} \longrightarrow [(H_2O)_4Zn^{II}(O_2)_2] \longrightarrow (H_2O)_4Zn^{II}(O_2) + O_2$$

$$(2\text{-}46)$$

Preliminary studies by stopped-flow spectrophotometry in MeCN with [Zn$^{II}$(DMU)₆](ClO₄)₂ (DMU, dimethylurea) indicate that the rate constant for the

Table 2-7  Voltammetric Reduction of 2.1 mM $O_2$ in the Presence of a 10- to 30-Fold Excess (Relative to $O_2$ Concentration) of Metal Cations in $Me_2SO$ (0.1 M TEAP)[a] at a Pt Electrode (0.02 V $s^{-1}$)

| Metal[b] | V versus NHE | $n(e^-/O_2)$ |
|---|---|---|
| – | -0.56 | 1 |
| $Li^+$ | -0.55 | 1 |
| $Zn^{2+}$ | -0.43 | 2 |
| $Cd^{2+}$ | -0.43 | 2 |
| $Fe^{2+}$ | -0.28 | 2 |
| $Mn^{2+}$ | -0.37 | 4 |
| $Co^{2+}$ | -0.42 | 4 |

[a] Tetraethylammonium perchlorate.

[b] Added as the hexahydrated perchlorate salts (Ref. 3).

reaction of Eq. (2-45) is greater than $10^8$ $M^{-1}$ $s^{-1}$. With slow scan rates and in the absence of "filming" of the electrode by $(H_2O)_4Zn^{II}(O_2)$, the overall process approaches the net reaction with an $O_2$ peak height that is equivalent to a two-electron process. However, in MeCN, DMF, and py there is a strong tendency for $Zn^{II}(O_2)$ to precipitate on the electrode surface and passivate it for the $O_2$ reduction process (see Fig. 2-8). Similar processes and problems undoubtedly occur with the other metal ions of Table 2-7 that promote the formation of insoluble or reactive metal peroxides.

$$Zn^{II}(OH_2)_4^{2+} + O_2 + 2\,e^- \longrightarrow (H_2O)_4Zn^{II}(O_2) \qquad (2\text{-}47)$$

In the case of Mn(II) or Co(II) the initial Lewis-acid-promoted disproportionation of $O_2^-$· is followed by further degradation steps to achieve an overall four-electron-per-$O_2$ process. For example,

$$(H_2O)_4Mn^{II}(O_2) + Mn^{II}(OH_2)_4^{2+} \longrightarrow [(HO)Mn^{III}OMn^{III}(OH)]^{2+} + 7\,H_2O \quad (2\text{-}48)$$

$$[(HO)Mn^{III}OMn^{III}(OH)]^{2+} + 2\,O_2^-· + H_2O \xrightarrow{\;H_2O\;} 2\,Mn^{II}(OH)_2(OH_2)_4 + 2\,O_2$$
$$(2\text{-}49)$$

These and related reactions lead to an overall reaction for the Mn(II)-catalyzed reduction of $O_2$

$$O_2 + 4\,Mn^{II}(OH_2)_4^{2+} + 2\,H_2O + 4\,e^- \longrightarrow 4\,(H_2O)_4^+Mn^{II}(OH) \quad (2\text{-}50)$$

Most of these metal–oxygen intermediates are expected to be reactive toward organic substrates and electrode surfaces. Hence, the presence of metal cations enhances the electron stoichiometry for the reduction of $O_2$, but frequently passivates the electrode surface. Thus, in the case of the formation of $(H_2O)_4Zn^{II}(O_2)$ on the surface of a platinum electrode, via the reactions of Eqs. (2-45) and (2-46), a likely process is a metathesis reaction

$$(H_2O)_4Zn^{II}(O_2) + Pt^{II}O \longrightarrow Pt^{II}(O_2) + Zn^{II}O + 4\,H_2 \qquad (2\text{-}51)$$

Likewise, a similar process may occur between the electrode surface and the HOOH that is produced via the reaction of Eq. (2-44).

$$Pt^{II}O + HOOH \longrightarrow Pt^{II}(O_2) + H_2O \qquad (2\text{-}52)$$

In the case of Mn(II) ions, the presence of HOOH from the proton-induced disproportionation of $O_2^{-\cdot}$ [Eq. (2-44)] can lead to the production of HO$\cdot$ radicals

$$(H_2O)_4^{2+}Mn^{II} + HOOH \longrightarrow (H_2O)_4^{2+}Mn^{III}(OH) + HO\cdot \qquad (2\text{-}53)$$

$$Pt + HO\cdot \longrightarrow [\cdot Pt^{I}(OH)] \xrightarrow{\;HO\cdot\;} Pt^{II}O + H_2O \qquad (2\text{-}54)$$

Because the weakest solvation of $O_2^{-\cdot}$ occurs in MeCN (of the solvents that have been considered in Table 2-5), superoxide should exhibit its maximum reactivity in this solvent. The peak separations for the $O_2/O_2^{-\cdot}$ cyclic voltammograms with metal electrodes strongly support this proposition (Fig. 2-9 and Table 2-5). To account for the extreme peak separation for the $O_2$ couple at platinum in MeCN, a reaction sequence that involves the formation of platinum peroxides is proposed

$$Pt^{II}O(Pt) + 2\,O_2^{-\cdot} \xrightarrow[2\,HO^-]{H_2O} 2\,Pt^{II}(O_2) \xrightarrow[H_2O]{2\,Pt^{II}O} 2\,Pt^{II}(OH)OOPt^{II}(OH) \qquad (2\text{-}55)$$

$$Pt^{II}(OH)OOPt^{II}(OH) + 2\,HO^- \longrightarrow 2\,Pt^{II}O + O_2 + 2\,H_2O + 2\,e^- \qquad (2\text{-}56)$$

The latter process would be expected to occur at a more positive potential than that for the oxidation of $O_2^{-\cdot}$. Processes similar to the reactions of Eqs. (2-55) and (2-56) have been observed with the electrochemical reduction of HOOH at mercury electrodes in DMF and MeCN. The almost reversible cyclic voltammogram for $O_2$ at a glassy carbon electrode in MeCN indicates that specific reactions occur between metal electrodes and $O_2^{-\cdot}$ in a poorly solvating medium for anions, such as MeCN.

*Electrode material effects*

Figure 2-9 illustrates the cyclic voltammograms for the reduction of $O_2$ in MeCN (0.1 $M$ TEAP) at glassy carbon, platinum, gold, and mercury electrodes.[3] The peak potentials and peak separations for the reduction of $O_2$ and reoxidation of $O_2^-$· with these electrodes in MeCN, Me$_2$SO, DMF, and py are summarized in Table 2-5.

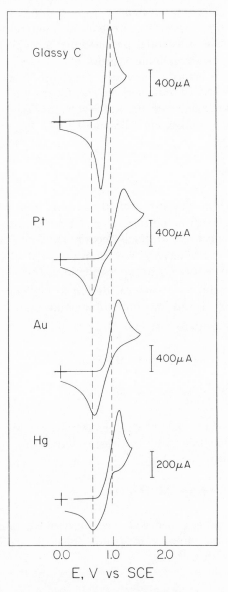

Figure 2-9 Cyclic voltammograms in MeCN (0.1 $M$ TEAP) of 8.1 m$M$ $O_2$ (at 1 atm) at vitreous carbon, platinum, gold, and mercury electrodes; electrode areas, 0.74 cm$^2$ (except mercury, 0.3 cm$^2$); scan rate 0.1 V s$^{-1}$

Clearly, the larger peak separations ($\Delta E_p$) for the metal electrodes, especially in MeCN, indicate that the apparent irreversibility for the $O_2/O_2^-$· couple is due to the reaction of $O_2^-$·, or its disproportionation product (HOO⁻), with the metal surface. This effect is largest in MeCN because it is the poorest solvating agent for $O_2^-$·, which is equivalent to minimizing its deactivation.

The significant effects of media and electrode material upon the reduction potentials and voltammetric peak currents for dissolved dioxygen require substantial knowledge of the media and appropriate calibration of the electrochemical electrode system before identification and quantitative determinations of $O_2$ are possible. Some of the interferences (e.g., $CH_3Cl$, $HCCl_3$, and $CCl_4$) may have substantial permeability through the membranes of the polarographic oxygen-membrane sensors (Clark electrode), which will cause substantial positive errors. However, these problems can be turned to an advantage. The enhanced reduction currents for the reduction of $O_2$ (at 1 atm) in the presence of alkyl halides, esters, and acyl halides can be used for their specific determination (with appropriate calibration).

*Chemical reduction*

In addition to its reduction by solvated electrons from pulse radiolysis, $O_2$ also can be reduced via electron transfer by several chemical agents. Because there are no apparent kinetic barriers to single-electron transfer to $O_2$, any reductant with a sufficiently negative one-electron reduction potential [($E_R^{o'}$) more negative than $E^{o'}O_2/O_2^-$·] should reduce $O_2$ to superoxide ion. This is believed to be the basis for the *in situ* generation of superoxide ion in biological matrices by reduced ferridoxin [$(Fe_4S_4)_R^{2-}$] and the semiquinonelike state of flavins (Fl⁻·) in the xanthine/xanthine oxidase system.[15,32]

$$(Fe_4S_4)_{red}^{2-} + O_2 \longrightarrow (Fe_4S_4)_{ox}^- + O_2^-· \qquad (2\text{-}57)$$

$$Fl^-· + O_2 \longrightarrow Fl_{ox} + O_2^-· \qquad (2\text{-}58)$$

Also, the introduction of methyl viologen ($MV^{2+}$, paraquat) in biological systems results in the formation of its one-electron reduction product, $MV^+$·, which mediates the production of $O_2^-$·

$$MV^+· + O_2 \longrightarrow MV^{2+} + O_2^-· \qquad (2\text{-}59)$$

In turn, the latter species is believed to be the cytotoxic agent that results from the application of methyl viologen on plants engaged in photosynthesis (photon-driven electron transfer).[33]

*Reduction by atom transfer*

In contrast to the preceding electron-transfer mechanism with inert electrodes,

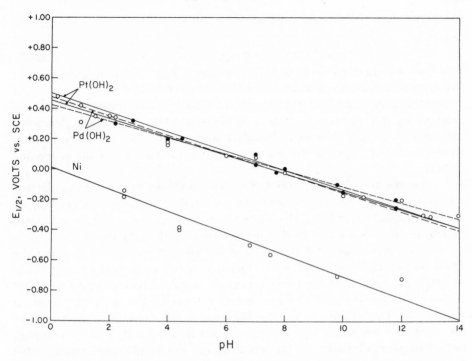

Figure 2-10 Voltammetric (slow-scan) data as a function of $p$H for the reduction (a) of freshly formed metal oxides [$Pt^{II}(OH)_2$ and $Pd^{II}(OH)_2$] (closed circles); and (b) of $O_2$ (1 atm) with preoxidized electrodes [$Pt^{II}(OH)_2$, $Pd^{II}(OH)_2$, $Ni^{II}(OH)_2$] (open circles). All solutions contained 0.1 $M$ $K_2SO_4$; $H_2SO_4$ and NaOH were used to adjust the $p$H.

the reduction of dioxygen at freshly preoxidized metal electrodes is $p$H dependent, yields negligible amounts of hydrogen peroxide, and occurs at the same potential as that for the metal oxide of the electrode material (see Fig. 2-10). A plausible explanation for these observations is represented by the reaction sequence

$$M^{II}(OH)_2 + 2\,H_3O^+ + 2\,e^- \longrightarrow M + 4\,H_2O \qquad (2\text{-}60a)$$

$$M^{II}(OH)_2 + 2\,e^- \longrightarrow M + 2\,HO^- \qquad (2\text{-}60b)$$

$$2\,M + O_2 + 2\,H_2O \longrightarrow 2\,M^{II}(OH)_2 \qquad (2\text{-}61)$$

In this mechanism, fresh metal oxide, $M^{II}(OH)_2$, is electrochemically reduced and reformed by a chemical oxidation of the surface by the dioxygen in the solution. The scheme satisfies the requirement that the reduction potential and rate of reaction be dependent on the electrode material, $M$, and $p$H. The net reaction is

$$O_2 + 2 H_2O + 4 e^- = 4 HO^-, \quad E_B^\circ, -0.40 \text{ V versus NHE} \tag{2-62}$$

or

$$O_2 + H_3O^+ + 4 e^- = 4 H_2O, \quad E^\circ = +1.23 \text{ V} \tag{2-63}$$

and does not involve formation of peroxide. Representative experimental reduction potentials at platinum are +0.05 V versus NHE and -0.65 V at $pH$ 2 and 13, respectively (Fig. 2-10). Although Eq. (2-60) clearly represents the rate-determining step in the reduction of dioxygen at preoxidized electrodes, some dioxygen may react directly at freshly reduced electrode sites to form trace amounts of peroxide via the electron-transfer mechanism of the preceding section.

Because the electron-transfer reduction of $O_2$ is a reversible one-electron process, the hope for an electron-transfer catalyst is futile. Atom-transfer catalysts [Eqs. (2-60) and (2-61)] can promote more extensive reduction and higher potentials (those that correspond to the reduction of the metal oxide).

The proposition of a one-electron mechanism for the electron-transfer reduction of dioxygen and the associated conclusions present significant ramifications relative to the development of improved fuel cells and metal–air batteries. To date the practical forms of such systems have used strongly acidic or basic electrolytes. Such solution conditions normally cause atom transfer to be the dominant reduction process for molecular oxygen at metal electrodes. Hence, the search for effective catalytic materials should be in this context rather than in terms of a one-electron-transfer process.

The third step of the scheme, whereby the metal is spontaneously oxidized to metal oxide [Eq. (2-61)] is the general equation for the corrosion of metals. This appears to involve the splitting of the oxygen–oxygen bond as oxygen atoms are transferred to the metal. While the mechanism of this reaction is unclear, extensive thermodynamic data are available for the oxidation of metals by molecular oxygen.

*Concerted one-electron reductions*

Reduction of $O_2$ in the presence of excess zinc cation [$Zn^{II}(bpy)_2^{2+}$], (tetraphenylporphinato)iron(III) ion [$(H_2O)^+Fe^{III}TPP$], and cuprous ion [$Cu^I(MeCN)_4^+$] results in formation of metal–dioxygen adducts. Figures 2-11 and 2-12 illustrate the cyclic voltammograms for ($Fe^{III}TPP$)Cl and [$Cu^I(MeCN)_4$]ClO$_4$, respectively, in the absence and in the presence of $O_2$. Reaction schemes for the three metal–$O_2$ systems are outlined:

(a)  $Zn^{II}(bipy)_2^{2+} + O_2 + e^- \xrightarrow{\;-0.5 \text{ V}\;} [Zn^{II}(bipy)_2OO\cdot]^+$ \hfill (2-64)

$[Zn^{II}(bipy)_2OO\cdot]^+ + e^- \longrightarrow Zn^{II}(bipy)_2(O_2)$ \hfill (2-65)

(b)  $Fe^{II}TPP + O_2 + e^- \xrightarrow[Me_2SO]{-0.5 \text{ V}} TPPFe^{III}OO^-$ \hfill (2-66)

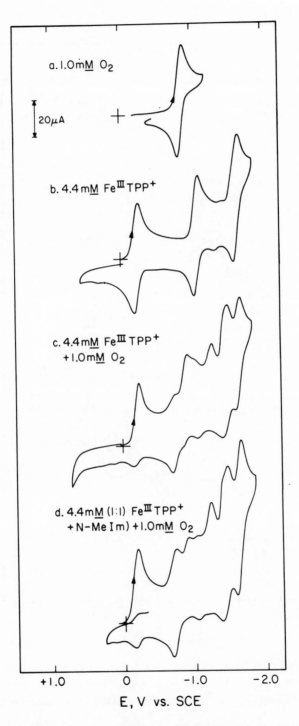

Figure 2-11 Cyclic voltammograms in DMF (0.1 $M$ TEAP) of (a) $O_2$, (b) $Fe^{III}$TPPCl, (c) $O_2$ + $Fe^{III}$TPPCl, and (d) $O_2$ + $Fe^{III}$TPP(N-MeIm)$^+$. Measurements were made with a glassy carbon electrode (area, 0.11 cm$^2$) at a scan rate of 0.1 V s$^{-1}$; temperature, 25°C.

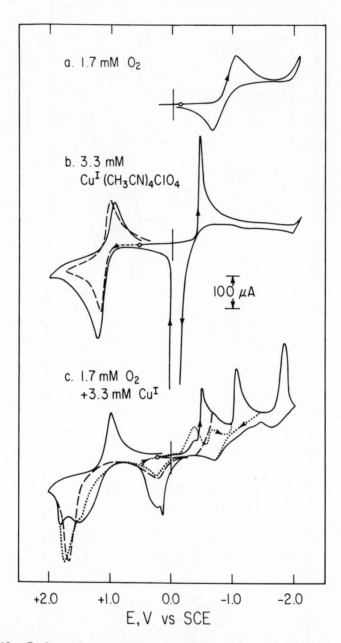

Figure 2-12  Cyclic voltammograms in MeCN (0.1 *M* TEAP) of (a) O$_2$, (b) [Cu$^I$(MeCN)$_4$]ClO$_4$, (c) O$_2$ + [Cu$^I$(MeCN)$_4$]ClO$_4$.  Measurements were made with a platinum electrode (area, 0.23 cm$^2$) at a scan rate of 0.1 V s$^{-1}$; temperature, 25°C.

$$Fe^{III}TPP^+ + O_2 + 2\,e^- \xrightarrow[\text{DMF, Me-Im}]{-0.55\ V} TPPFe^{III}OO^- \xrightarrow[2\,x]{H_2O}$$

$$(TPP)Fe^{III}\text{-}O\text{-}Fe^{III}(TPP) + HOOH + O_2 + 2\ HO^- \qquad (2\text{-}67)$$

(c) $\quad [Cu^I(MeCN)_4]ClO_4 + O_2 + e^- \xrightarrow{-0.35\ V} Cu^I(O_2) \qquad\qquad (2\text{-}68)$

$$Cu^I(O_2) + e^- \xrightarrow{-1.35\ V} Cu^I(O_2)^- \xrightarrow[H_2O]{e} \cdot Cu^{II}O + 2\ HO^- \qquad (2\text{-}69)$$

$$\cdot Cu^{II}O + e^- \xrightarrow[H_2O]{-1.8\ V} Cu^I(OH)_2^- \qquad\qquad (2\text{-}70)$$

$$Cu^I(O_2) \xrightarrow{+2.1\ V} \cdot Cu^{II}(MeCN)_4^{2+} + O_2 + 2\,e^- \qquad\qquad (2\text{-}71)$$

$$2\ Cu^I(MeCN)_4^+ + O_2 + 2\,e^- \xrightarrow{-0.25\ V}$$

$$Cu^IOOCu^I \longrightarrow 2\cdot Cu^{II}O \qquad\qquad (2\text{-}72)$$

$$Cu^IOOCu^I \underset{-0.05\ V}{\overset{+0.55\ V}{\rightleftharpoons}} [CuOOCu]^{2+} + 2\,e^- \qquad\qquad (2\text{-}73)$$

When the metal-ion cofactor is electroactive at potentials equal or more positive than that for the one-electron reduction of $O_2$, then an effective two-electron reduction is accomplished by two sequential one-electron reductions [see Eq. (2-67)]. Such a process is similar to the favored $O_2$-activation mechanism for cytochrome *P*-450.[34]

## Applications

At one time the main route for the production of hydrogen peroxide was the electrolytic reduction of $O_2$ in acidic media

$$O_2 + 2\ H_3O^+ + 2\,e^- \longrightarrow HOOH \qquad\qquad (2\text{-}74)$$

Although current industrial production is via chemical reduction, the electrolysis process is convenient for small quantities of HOOH.

The main application of dioxygen electrochemistry has been and continues to be in the realm of analytical chemistry. The preceding sections of this chapter provide ample evidence that for carefully controlled conditions the current $(i_{O_2})$ for the reduction of $O_2$ is directly proportional to its concentration [which is directly proportional to its partial pressure $(P_{O_2})$]

$$i_{O_2} = k\,[O_2] = k'P_{O_2} \tag{2-75}$$

Although this has been known since the very beginning of electrochemistry in the nineteenth century, commercial electroanalytical instrumentation has been available only for the past 32 years. The direct reduction of $O_2$ in aqueous solutions (especially biological and water treatment/sewage samples) is subject to positive interference by electroactive components of the sample (transition metals, oxygenated organic molecules, peroxides) and negative interference from organic matter, proteins, lipids, and carbohydrates in the sample, which film the electrode. All of these problems were alleviated by separating the sample solution (or gas sample) and the electrode surface (and the electrolysis cell) with a gas-permeable membrane (1-mm polyethylene or Teflon). This ingenious advance was discovered in 1954 by Leland C. Clark, Jr., and is the basis for dioxygen sensors and analytical instrumentation for biological fluids, sewage and water treatment plants, and gas mixtures.[35,36] Because the polymer membrane constitutes the limit for $O_2$ diffusion to the electrode, the Clark electrode gives a current response that is proportional to the partial pressure of $O_2$ ($P_{O_2}$). Thus, it will give an identical signal for a pure $O_2$ gas sample (1 atm, 44 m$M$) and a aqueous solution saturated with $O_2$ at 1 atm ($O_2$ concentration, 1 m$M$).

Another analytical sensor that is based on the electrolytic reduction of $O_2$ has been commercialized during the past decade. This is a device for the rapid, accurate assay of glucose in human blood and makes use of an immobilized enzyme (glucose oxidase) on a disposable sample probe that includes an $O_2$-electrolysis sensor with electrolyte, buffer, and various inhibitors. When a drop of blood is placed on the sample probe, the glucose oxidase (GO) transformation of the glucose begins via the reduction of the ambient $O_2$ to HOOH.

$$O_2 + D\text{-glucose} \xrightarrow{\quad GO \quad} HOOH + d\text{-gluconic acid} \tag{2-76}$$

The most elegant and effective form of this sensor uses ferricenium ion $[Fe^{III}(Cp)_2^+$ in place of $O_2$ as the election acceptor.[37] The electrolysis current of the sensor, which is proportional to the $Fe^{III}(Cp)_2^+$ concentration, decreases as the enzymatic reaction [Eq. (2-76)] consumes it. The rate of change in the current for $Fe^{III}(Cp)_2^+$ reduction is proportional to the $D$-glucose concentration, and is the basis of the assay.

$$\frac{-d[Fe^{III}(Cp)_2^+]}{dt} = k\,[Fe^{III}(Cp)_2^+]\,[\text{glucose}]\,[GO] \tag{2-77}$$

$$\frac{-d(\ln i_{Fe}{}^+)}{dt} = \frac{-d(\ln i_{Fe}{}^+)}{i_{Fe}{}^+\,dt} = \frac{-d[Fe^{III}(Cp)_2^+]}{[Fe^{III}(Cp)_2^+]\,dt} = k'''\,[\text{glucose}] \tag{2-78}$$

Thus, any oxidase–substrate reaction that consumes $O_2$ (ferricenium ion) in a well-defined process is amendable to the development of electrochemical sensors and assays for substrates and enzyme activity.

*Reduction of HOOH*

The electrochemical reduction of HOOH in pyridine yields superoxide ion.[38] That a reduction process generates a species with a higher oxidation state than HOOH is surprising. However, the cyclic voltammogram for the reduction of HOOH in pyridine at a platinum working electrode exhibits a broad cathodic peak at -0.95 V versus NHE and anodic peaks for the reverse scan at -0.50 and -0.15 V versus NHE. The latter is characteristic of electrolytically generated $H_2$. Controlled-potential reduction of HOOH in pyridine at -1.0 V versus NHE (with argon degassing) results in a solution that exhibits an anodic cyclic voltammogram that is characteristic of $O_2^-$. ESR studies of the reduced solution confirm the presence of superoxide ion.

The products and the observed electron stoichiometries for the electrochemical reduction of HOOH are consistent with a mechanism in which the primary step is a one-electron transfer,

$$HOOH + e^- \longrightarrow HOO^- + \tfrac{1}{2}H_2, \quad E^{\circ\prime}, \ -1.0 \text{ V versus NHE} \qquad (2\text{-}79)$$

followed by a chemical reaction with another hydrogen peroxide molecule. The resultant $HO\cdot$ is trapped by the pyridine solvent to yield a stable solution of $O_2^-$ [Eq. (2-80)].[39]

$$HOOH + HOO^- \longrightarrow O_2^- + H_2O + HO\cdot$$
$$\xrightarrow[k,\, 3 \times 10^9 \, M^{-1} s^{-1}]{py} \ \tfrac{1}{n}[py(OH)]_n \qquad (2\text{-}80)$$

In acetonitrile there is no evidence of superoxide ion, either by ESR or by cyclic voltammetry, from the electrochemical reduction of HOOH. This can be explained by the slow rate of reaction of $HO\cdot$ with $CH_3CN$ ($k$, $4 \times 10^6 \, M^{-1} s^{-1}$), which permits $HO\cdot$ and $O_2^-$ to react to give $O_2$.

$$HO\cdot + O_2^- \xrightarrow{k,\, 1 \times 10^{10} \, M^{-1} s^{-1}} O_2 + HO^- \qquad (2\text{-}81)$$

The reduction of hydrogen peroxide in aqueous solution appears to be analogous to that in $CH_3CN$, with the mechanism represented by the reaction of Eq. (2-79), followed by the reactions of Eqs. (2-80) and (2-81). Thus, the reduction of HOOH yields $H_2$ and $HOO^-$ initially, in a one-electron step. The final products are the result of the reaction of $HOO^-$ and HOOH, and are analogous to those for the base-induced decomposition of HOOH.[39]

*Oxidation of HOOH and HOO⁻*

In acetonitrile HOOH is oxidized to $O_2$ via an electron-transfer/chemical/ electron-transfer (ECE) mechanism

$$\text{HOOH} \xrightarrow{-e^-} [\text{HOOH}]^{+\cdot} \xrightarrow{\text{H}_2\text{O}} \text{HOO}\cdot + \text{H}_3\text{O}^+, \quad E^{o\prime}, + 2.0 \text{ V versus NHE}$$
$$(2\text{-}82)$$

$$\text{HOO}\cdot \xrightarrow[\text{H}_2\text{O}]{-e^-} \text{O}_2 + \text{H}_3\text{O}^+, \quad E^\circ, + 0.6 \text{ V versus NHE} \qquad (2\text{-}83)$$

Although HOO⁻ reacts rapidly with most organic solvents, it persists long enough in pyridine to permit its electrochemical oxidation via a similar ECE mechanism.[40]

$$\text{HOO}^- \xrightarrow{-e^-} [\text{HOO}\cdot] \xrightarrow{\text{HO}^-} \text{H}_2\text{O} + \text{O}_2^{-\cdot}, \quad E^\circ, -0.37 \text{ V versus NHE} \qquad (2\text{-}84)$$

$$\text{O}_2^{-\cdot} \xrightarrow{-e^-} \text{O}_2, \quad E^\circ, -0.66 \text{ V versus NHE} \qquad (2\text{-}85)$$

## Summary

The redox chemistry of dioxygen and its reduction products is heavily dependent upon mechanistic pathway, substrate, and solution acidity.  For those circumstances that are limited by direct electron transfer, the redox mechanisms for dioxygen species involve one-electron steps.  To achieve the full oxidizing potential and energetics for $O_2$ requires catalysts that facilitate atom-transfer mechanisms and/or stabilize dioxygen intermediates.  The dominant characteristic of $O_2^{-\cdot}$ (the primary product from the electron-transfer reduction of $O_2$) is its ability to act as a Brønsted base.  Under aprotic conditions $O_2^{-\cdot}$ is a strong nucleophile, a moderate one-electron reducing agent, a source of $^1O_2$ when oxidized by a strong one-electron oxidant with a closed coordination sphere, and a hydrogen-atom-transfer oxidant of basic substrates such as hydroxylamines and reduced flavins.  The base-induced disproportionation of HOOH leads to the formation of $O_2^{-\cdot}$ and HO· in aprotic media.  The variety of oxidation-reduction, hydrolytic, atom-transfer, and dismutation reaction pathways for dioxygen species is ample evidence of the unique chemical versatility of $O_2$ and HOOH.

## References

1. Metzner, H. (ed.). *Photosynthetic Oxygen Evolution*. New York: Academic Press, 1978.

2. Barrette, W. C., Jr.; Johnson, H. W., Jr.; Sawyer, D. T. *Anal. Chem.* **1984,** *56,* 1890.

3. Sawyer, D. T.; Chiericato, G., Jr.; Angelis, C. T.; Nanni, E. J., Jr.; Tsuchiya, T. *Anal. Chem.* **1982,** *54,* 1720.

4. Parsons, R. *Handbook of Electrochemical Constants*. London: Butterworths Scientific Publications, 1959.

5. Schwarz, H. A.; Dodson, R. W. *J. Phys. Chem.* **1984,** *88,* 3643.

6 Wilshire, J.; Sawyer, D. T. *Accts. Chem. Res.* **1979,** *12,* 105.

7. Bard, A. J.; Parsons, R.; Jordon, J. *Standard Potentials in Aqueous Solution*. New York: Marcel Dekker, 1985.

8. Cofré, P.; Sawyer, D. T. *Anal. Chem.* **1986**, *58*, 1057.

9. Cofré, P.; Sawyer, D. T. *Inorg. Chem.* **1986**, *25*, 2089.

10. Sawyer, D. T.; Roberts, J. L.; Jr.; Tsuchiya, T.; Srivatsa, G. S.; In *Oxygen Radicals in Chemistry and Biology* (Bors, W.; Saran, M.; Tait, D., eds.). Berlin: Walter de Gruyter and Co., 1984.

11. Roberts, J. L. Jr.; Morrison, M. M.; Sawyer, D. T. *J. Am. Chem. Soc.* **1978**, *100*, 329.

12. Roberts, J. L., Jr.; Sawyer, D. T. *Israel J. Chem.* **1983**, *23*, 430.

13. Chin, D.-H.; Chiericato, G. Jr.; Nanni, E. J., Jr.; Sawyer, D. T. *J. Am. Chem. Soc.* **1982**, *104*, 1296.

14. Gibian, M. J.; Sawyer, D. T.; Ungerman, T.; Tangpoonpholvivat, R.; Morrison, M. M. *J. Am. Chem. Soc.* **1979**, *101*, 640.

15. Fridovich, I. In *Oxygen and Oxy Radicals* (Rodgers, M. A. J.; Powers, E. L., eds.). New York: Academic Press, 1981.

16. Fee, J. A. In *Oxygen and Oxy Radicals* (Rodgers, M. A. J.; Powers, E. L., eds.) New York: Academic Press, 1981.

17. Sawyer, D. T.; McDowell, M. S.; Yamaguchi, K. S., *Chem. Res. Toxicol.* **1988**, *1*, 97.

18. Tsang, P. K. S.; Cofré, P.; Sawyer, D. T. *Inorg. Chem.* **1987**, *26*, 3604.

19. Roberts, J. L.; Jr.; Sawyer, D. T. *J. Am. Chem. Soc.* **1981**, *103*, 712.

20. Roberts, J. L.; Jr.; Calderwood, T. S.; Sawyer, D. T. *J. Am. Chem. Soc.* **1983**, *105*, 7691.

21. Laitinen, H. A.; Kolthoff, I. M. *J. Phys. Chem.* **1941**, *45*, 1061.

22. Sawyer, D. T.; Interrante, L. V. *J. Electroanal. Chem.* **1961**, *2*, 310.

23. Sawyer, D. T.; Day, R. J. *Electrochim. Acta* **1963**, *8*, 589.

24. Bielski, B. H. J. *Photochem. Photobiol.* **1978**, *28*, 645.

25. Bielski, B. H. J.; Allen, A. O. *J. Phys. Chem.* **1977**, *81*, 1048.

26. Ilan, Y. A.; Meisel, D.; Czapski, G. *Israel J. Chem.* **1974**, *12*, 891.

27. Behar, D.; Czapski, G.; Rabini, J.; Dorfman, L. M.; Schwartz, H. A. *J. Phys. Chem.* **1970**, *74*, 3209.

28. Rabini, J.; Nielsen, S. O. *J. Phys. Chem.* **1969**, *73*, 3736.

29. Sawyer, D. T.; Roberts, J. L., Jr. *J. Electroanal. Chem.* **1966**, *12*, 90.

30. Sawyer, D. T.; Seo, E. T. *Inorg. Chem.* **1977**, *16*, 499.

31. Roberts, J. L., Jr.; Morrison, M. M.; Sawyer, D. T. *J. Am. Chem. Soc.* **1978**, *100*, 329.

32. McCord, J. M.; Fridovich, I. *J. Biol. Chem.* **1968**, *243*, 5753.

33. Nanni, E. J., Jr.; Angelis, C. T.; Dickson, J.; Sawyer, D. T. *J. Am. Chem. Soc.* **1981**, *103*, 4268.

34. Ortiz de Montellano, P. R. (ed.). *Cytochrome P-450: Structure, Mechanism, and Biochemistry.* New York: Plenum, 1986.

35. Clarke, L. C., Jr., *Trans. Am. Soc. Artif. Intern. Organs* **1956**, *2*, 41.

36. Sawyer, D. T. and J. L., Roberts, Jr. *Electrochemistry for Chemists.* New York: Wiley-Interscience, 1974, pp. 383–384.

37. Cass, A. E. G.; Davis, G.; Francis, G. D.; Hill, H. A. O., Aston, W. J.; Higgins, I. J.; Plotkin, E. V.; Scott, L. D. L.; Turner, A. P. F. *Anal. Chem.* **1984**, *56*, 667.

38. Morrison, M. M.; Roberts, J. L., Jr.; Sawyer, D. T. *Inorg. Chem.* **1979**, *18*, 1971.

39. Roberts, J. L., Jr.; Morrison, M. M.; Sawyer, D. T. *J. Am. Chem. Soc.* **1978**, *100*, 329.

40. Tsang, P. K. S.; Jeon, S.; Sawyer, D. T., *Inorg. Chem.*, submitted, 1991.

# 3

# NATURE OF CHEMICAL BONDS FOR OXYGEN IN ITS COMPOUNDS

*Molecules:   Arrays of atoms connected by covalent bonds*

The bonding for oxygen atoms in heteratomic molecules is viewed as essentially covalent [e.g., MeOH, $Me_2C=O$, MeCH(O), and MeC(O)OOH] and similar to that for carbon, nitrogen, and chlorine atoms.   In contrast, a Lewis acid–base formalism often is used for metal–oxygen compounds with ionic interactions by dianionic oxo groups (e.g., $[Ba^{2+}O^{2-}]$, $[Fe^{6+}(O^{2-})_4]^{2-}$, $[Mn^{7+}(O^{2-})_4]^-$, and $[(Cu^+)_2O^{2-}]$). This results from the thermodynamic relations for ionic solution equilibria, and the inference to some that the combination of ions results in molecules and complexes held together by electrostatic interactions, for example,

$$H_3O^+ + HO^- \rightleftharpoons 2\,H_2O, \; K_f, 10^{14} \tag{3-1}$$

$$Zn^{2+} + 4\,HO^- \rightleftharpoons Zn(OH)_4^{2-}, \; K_f, 10^1 \tag{3-2}$$

$$Fe^{3+} + 3\,HO^- \rightleftharpoons Fe(OH)_3(s), \; K_f, 10^{36} \tag{3-3}$$

$$H_3O^+ + Cl^- \overset{H_2O}{\rightleftharpoons} HCl + H_2O, \; K_f, 10^{-4} \tag{3-4}$$

$$H^+ + Cl^- \overset{MeCN}{\rightleftharpoons} HCl, \; K_f, 10^4 \tag{3-5}$$

$$Ag^+ + Cl^- \overset{H_2O}{\rightleftharpoons} AgCl(s), \; K_f, 10^{10} \tag{3-6}$$

However, the bonding in the $H_2O$ molecule is essentially covalent [H–OH ($1s$-$2p$)] with a charge excess of 0.12+ per hydrogen atom (0.25- for oxygen), and *not* [$H^+$ $^-OH$].[1,2]   The same is true for all of the molecular products in Eqs. (3-1) – (3-6).

Thus, the bonding in metal compounds and complexes has traditionally been viewed as ionic (positive metal center interacting with negative ions; $HO^-$, $O^{2-}$, $Cl^-$, $AcO^-$) and coordinate donor [Lewis acid–base interactions; positive metal center interacting with negative ions and electron-pair Lewis bases (:NH₃, :PPh₃, HÖH)].   Examples of ionic versus covalent bonding illustrate the tradition; $H^+Cl^-$ versus   H–Cl ($1s$-$3p$),   $C^{4+}(Cl^-)_4$ versus   C(–Cl)₄ [$2sp^3$-$(3p)_4$],   $Fe^{3+}(Cl^-)_3$   versus Fe(–Cl)₃ [$3d^5sp^2$-$(3p)_3$], $H^+$ $^-OH$ versus H–OH ($1s$-$2p$), $C^{4+}(O^{2-})_2$ versus O=C=O

$[2sp^3=(2p^2)_2]$, and $Fe^{2+}(O^{2-})$ versus $Fe=O$ $[3d^6sp=(2p^2)]$. Such ionic formulations for these molecules in an inert matrix are not consistent with their physical and chemical properties. For example, ferrate ion $(FeO_4^{2-})$ is formulated as an ionic salt $Fe^{6+}(O_2^{2-})_4$ $(3d^2, S=2/2,$ tetrahedral). This, in spite of the fact that $O^{2-}$ is unattainable with electron beams in a vacuum, which would make it a stronger reducing agent than the electron (electrons transform all oxidized iron to $Fe^0$ or $Fe^-$). The continued propensity to use ionic bonds with 8-valence-electron oxygen [oxo, oxide, $O^{2-}$ $(2s^2p^6)$] in the formulations of metal–oxygen compounds is akin to our fondness for an emotional function for the human-heart muscle.

### Electronegativities of the elements and valence-bond theory

The fundamental axiom of the valence-bond theory[3] is that elemental atoms that are connected by covalent bonds are as close to electrical neutrality as is consistent with their respective electronegativities. Hence, uncharged molecules contain uncharged atoms that are held together by covalent bonds and react with each other (in an inert matrix) to give uncharged molecules (with uncharged atoms held together by covalent bonds). For example,

$(Na_2)(g)$ (bond energy, 17.6 kcal mol$^{-1}$) + $(Cl_2)(g)$ (BE, 59.0 kcal mol$^{-1}$)

$$\rightarrow 2\ (NaCl)(g)\ (BE,\ 98.5\ kcal\ mol^{-1}) \tag{3-7}$$

with a charge excess ($\delta$) of 0.33 e$^-$ for the chlorine of NaCl on the basis of its gas-phase dipole moment[4] as well as the difference in the electronegativities of Na and Cl (Table 3-1).[5–7]

The second axiom of the valence-bond theory is that chemical bonds in metal salts and complexes are a combination of covalent (electron-pair) and ionic (electrostatic positive-ion–negative-ion attraction) bonding.[8] Thus, for the HCl molecule ($\Delta H_{DBE}$, 103 kcal mol$^{-1}$)[9] the covalent portion of the bond energy is assumed to be equal to the geometric mean of the covalent bond energies for $H_2$ ($\Delta H_{DBE}$, 104 kcal mol$^{-1}$)[6] and $Cl_2$ ($\Delta H_{DBE}$, 59 kcal mol$^{-1}$),[9] $(104 \times 59)^{1/2}$ = 78 kcal mol$^{-1}$. The difference, $\Delta$ = (103-78) = 25 kcal mol$^{-1}$, has been assigned to electrostatic interactions between $H^{\delta+}$ and $Cl^{\delta-}$ of the HCl molecule (ionic portion of the bond energy), and is the basis of the Pauling scale of electronegativities,[3] $(\Delta/96) = (\chi_{Cl} - \chi_H)^2$, and the assumption that HCl(g) is 24% ionic $[(25/103)100 = 24]$. Because this is a circular argument with respect to an estimate of the degree of ionic bonding (and predicts that the bonding of CO is 49% ionic), a more reasonable approach is to use the Mulliken definition of electronegativity, which is one-half of the sum of the element's first ionization potential (IP) and its electron affinity (EA), $\chi$ = (IP + EA)/2.[8]

With modern data[5] the electronegativity values can be used to estimate the charge transfer ($\delta$ or $\Delta N/2$; $\Delta N$, differential charge of dipole) in diatomic molecules, as illustrated in Table 3-2 for HCl, metal–chlorine molecules, and metal–oxygen molecules. The charge excesses ($\delta$) for the chlorine and oxygen of several diatomic molecules [estimated on the basis of their respective

Table 3-1 Electronegativities of the Elements [$\chi$ = (IP + EA)/2][a-c]

Elements[e]

| Group | I | II | | | | | | | | | | | III | IV | V | VI | VII | VIII |
|---|---|---|---|---|---|---|---|---|---|---|---|---|---|---|---|---|---|---|
| | H<br>13.60[a]<br>0.75[b]<br>7.18[c]<br>6.42[d] | | | | | | | | | | | | | | | | | He<br>24.59 |
| | Li<br>5.39<br>0.62<br>2.99<br>2.38 | Be<br>9.32<br>0.0<br>4.66<br>4.66 | | | | | | | | | | | B<br>8.30<br>0.28<br>4.29<br>4.01 | C<br>11.26<br>1.26<br>6.27<br>5.00 | N<br>14.53<br>0.0<br>7.26<br>7.26 | O<br>13.62<br>1.46<br>7.55<br>6.08 | F<br>17.42<br>3.40<br>10.41<br>7.01 | Ne<br>21.56 |
| | Na<br>5.14<br>0.55<br>2.84<br>2.30 | Mg<br>7.65<br>0.0<br>3.83<br>3.83 | | | | | | | | | | | Al<br>5.99<br>0.44<br>3.20<br>2.78 | Si<br>8.15<br>1.38<br>4.76<br>3.39 | P<br>10.49<br>0.75<br>5.62<br>4.87 | S<br>10.36<br>2.08<br>6.22<br>4.14 | Cl<br>12.97<br>3.62<br>8.30<br>4.68 | Ar<br>15.76 |
| | K<br>4.34<br>0.50<br>2.42<br>1.92 | Ca<br>6.11<br>0.0<br>3.04<br>3.04 | Sc<br>6.54<br>0.19<br>3.36<br>3.18 | Ti<br>6.82<br>0.08<br>3.46<br>3.37 | V<br>6.74<br>0.52<br>3.64<br>3.11 | Cr<br>6.77<br>0.67<br>3.72<br>3.05 | Mn<br>7.44<br>0.0<br>3.72<br>3.72 | Fe<br>7.87<br>0.15<br>4.01<br>3.86 | Co<br>7.86<br>0.61<br>4.27<br>3.63 | Ni<br>7.64<br>1.16<br>4.40<br>3.23 | Cu<br>7.73<br>1.24<br>4.48<br>3.25 | Zn<br>9.39<br>0.0<br>4.68<br>4.68 | Ga<br>6.00<br>0.3<br>3.15<br>2.85 | Ge<br>7.90<br>1.23<br>4.55<br>3.34 | As<br>9.81<br>0.81<br>5.31<br>4.50 | Se<br>9.75<br>2.02<br>5.88<br>3.87 | Br<br>11.81<br>3.36<br>7.60<br>4.22 | Kr<br>14.00 |
| | Rb<br>4.18<br>0.49<br>2.34<br>1.85 | Sr<br>5.70<br>0.0<br>2.86<br>2.86 | Y<br>6.38<br>0.31<br>3.33<br>3.04 | Zr<br>6.84<br>0.43<br>3.64<br>3.21 | Nb<br>6.88<br>0.89<br>3.88<br>3.00 | Mo<br>7.10<br>0.75<br>3.93<br>3.18 | Tc<br>7.28<br>0.55<br>3.90<br>3.37 | Ru<br>7.37<br>1.05<br>4.22<br>3.16 | Rh<br>7.46<br>1.14<br>4.29<br>3.16 | Pd<br>8.34<br>0.56<br>4.45<br>3.89 | Ag<br>7.58<br>1.30<br>4.45<br>3.14 | Cd<br>8.99<br>0.0<br>4.50<br>4.50 | In<br>5.79<br>0.30<br>3.04<br>2.75 | Sn<br>7.34<br>1.11<br>4.22<br>3.12 | Sb<br>8.64<br>1.07<br>4.87<br>3.79 | Te<br>9.01<br>1.97<br>5.49<br>3.52 | I<br>10.45<br>3.06<br>6.77<br>3.70 | Xe<br>12.13 |

| Cs | Ba | La[f] | Hf | Ta | W | Re | Os | Ir | Pt | Au | Hg | Tl | Pb | Bi | Po | At | Rn |
|------|------|------|------|------|------|------|------|------|------|------|------|------|------|------|------|------|------|
| 3.89 | 5.21 | 5.58 | 7.0 | 7.89 | 7.98 | 7.88 | 8.7 | 9.1 | 9.0 | 9.22 | 10.44 | 6.11 | 7.42 | 7.29 | 8.42 | – | 10.75 |
| 0.47 | 0.0 | 0.5 | ~0 | 0.32 | 0.81 | 0.15 | 1.10 | 1.56 | 2.13 | 2.31 | 0.0 | 0.2 | 0.36 | 0.95 | 1.9 | 2.8 | |
| 2.19 | 2.60 | 3.04 | 3.5 | 4.11 | 4.40 | 4.9 | 5.3 | 5.6 | 5.6 | 5.78 | 5.23 | 3.2 | 3.88 | 4.4 | 5.16 | | |
| 1.71 | 2.60 | 2.52 | 3.5 | 3.79 | 3.59 | 3.87 | 3.8 | 3.8 | 3.4 | 3.46 | 5.23 | 3.0 | 3.58 | 3.17 | 3.26 | | |

| Fr | Ra | Ac[g] |
|------|------|------|
| | 5.28 | 6.9 |
| | 0.0 | |
| | 2.64 | |
| | 2.64 | |

Lanthanide series

| Ce | Pr | Nd | Pm | Sm | Eu | Gd | Tb | Dy | Ho | Er | Tm | Yb | Lu |
|------|------|------|------|------|------|------|------|------|------|------|------|------|------|
| 5.47 | 5.42 | 5.49 | 5.55 | 5.63 | 5.67 | 6.14 | 5.85 | 5.93 | 6.02 | 6.10 | 6.18 | 6.25 | 5.43 |

Actinide series

| Th | Pa | U | Np | Pu | Am | Cm | Bk | Cf | Es | Fm | Md | No | Lr |
|------|------|------|------|------|------|------|------|------|------|------|------|------|------|
| | | | | 5.8 | 6.0 | | | | | | | | |

[a] Ionization potential in electron volts (Ref. 5).

[b] Electron affinity in electron volts (Ref. 5).

[c] Electronegativity; $\chi = (IP+EA)/2$.

[d] Hardness; $\eta = (IP - EA)/2$.

[e] Electron transfer in XY; $\delta = (\chi_X - \chi_Y)/4(\eta_X + \eta_Y)$ and $\delta = |(\chi_X \chi_Y)^{1/2} - \chi_Y|/10.4$.

[f] Lanthanide.

[g] Actinide series.

Table 3-2  Bond Energies ($\Delta H_{DBE}$) and Charge Transfer [$\delta = \Delta N_{MY}/2$] for Diatomic Metal–Chlorine and Metal–Oxygen Molecules

| Metal | M–Cl Bond energy, [a] $\Delta H_{DBE}$ (kcal mol$^{-1}$) | $\delta$ [b] (Cl$^{-\delta}$) | M–O Bond energy, [a] $\Delta H_{DBE}$ (kcal mol$^{-1}$) | $\delta$ [b] (O$^{-\delta}$) | Valence-electron hybridization of M |
|---|---|---|---|---|---|
| H | 103±1 | 0.06 (0.09)[c] | 102±1 | | $s$ |
| Na | 98±2 | 0.33 (0.33) | 96±2 | | $s$ |
| Cs | 107±2 | 0.39 (0.31) | 105±2 | | $s$ |
| Ba | 104±2 | 0.35 | 134±3 | 0.30 (0.31)[c] | $sp$ |
| Zn | 55±5 | 0.20 | 65±10 | 0.15 | $d^{10}sp$ |
| Mn | 86±2 | 0.26 | 96±10 | 0.22 | $d^{5}sp$ |
| Fe | 84 | 0.24 | 93±4 | 0.20 | $d^{6}sp$ |
| Co | 93 | 0.23 | 92±3 | 0.18 | $d^{7}sp$ |
| Ni | 89±5 | 0.22 | 91±4 | 0.17 | $d^{8}sp$ |
| Cu | 92±1 | 0.21 | 64±5 | 0.17 | $d^{10}s$ ($d^{9}sp$ for CuO) |

[a] Dissociative bond energy for M–Cl or M–O; kcal mol$^{-1}$ (Ref. 9).

[b] $\delta = [\Delta N_{MY}/2] = [(\chi_M \times \chi_Y)^{1/2}/2 - \chi_Y]/10.4$ (Ref. 7); electronegativity, $\chi = (IP + EA)/2$ (Table 3-1).

[c] For HCl, the dipole moment is 1.08 debyes and the interatomic distance is 1.267 Å; $\Delta N_{HCl} = [+\delta-(-\delta)] = 2\delta = (1.08 \times 10^{-18})/(1.27 \times 10^{-8} \times 4.8 \times 10^{-10}) = 0.18$ (Ref. 4). For BaO, the dipole moment is 7.95 debyes and the interatomic distance is 2.66Å; $\Delta N_{BaO} = 2\delta = (7.95 \times 10^{-18})/(2.66 \times 10^{-8} \times 4.8 \times 10^{-10}) = 0.62$ (Ref. 4).

electronegativities $(\chi)]^{5-7}$ are summarized in Table 3-2 together with their gas-phase covalent bond energies ($\Delta H_{DBE}$). The gas-phase dipole moments for HCl, CsCl, KCl, and BaO and their interatomic distances provide a direct measure of the metal–chlorine and metal–oxygen charge transfer ($\Delta N$). The use of Mulliken electronegatives to predict $\Delta N$ appears consistent with the measured quantities, although the degree of charge transfer is overestimated by up to 18% for CsCl. In any case, the electronegativity differences for CsCl and BaO are the largest of the Periodic Table (if F is treated as a special case), yet their bonds are 70% covalent. Hence, to a first approximation all chemical bonds have substantial covalency, and neutral molecules are assembled from neutral atoms (with valence-electron hybridization to give unpaired electrons) via electron-pair formation (covalent bonds). For example,

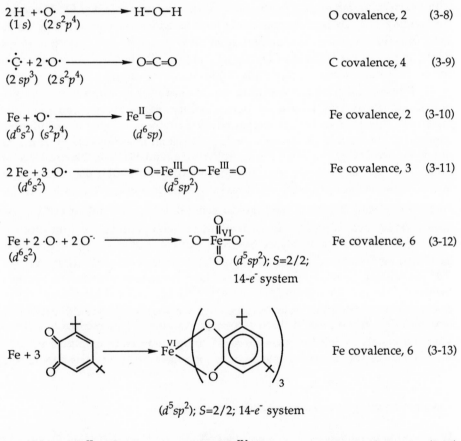

$$2\,H + \cdot \ddot{O} \cdot \longrightarrow H\text{–}O\text{–}H \qquad\qquad \text{O covalence, 2} \qquad (3\text{-}8)$$
$$(1\,s) \quad (2\,s^2 p^4)$$

$$\cdot \ddot{C} \cdot + 2\,\cdot \ddot{O} \cdot \longrightarrow O{=}C{=}O \qquad\qquad \text{C covalence, 4} \qquad (3\text{-}9)$$
$$(2\,sp^3) \quad (2\,s^2 p^4)$$

$$Fe + \cdot \ddot{O} \cdot \longrightarrow Fe^{II}{=}O \qquad\qquad \text{Fe covalence, 2} \qquad (3\text{-}10)$$
$$(d^6 s^2) \; (s^2 p^4) \qquad\qquad (d^6 sp)$$

$$2\,Fe + 3\,\cdot \ddot{O} \cdot \longrightarrow O{=}Fe^{III}\text{–}O\text{–}Fe^{III}{=}O \qquad\qquad \text{Fe covalence, 3} \qquad (3\text{-}11)$$
$$(d^6 s^2) \qquad\qquad\qquad (d^5 sp^2)$$

$$Fe + 2\,\cdot \ddot{O} \cdot + 2\,O^{-} \longrightarrow {}^{-}O\text{–}\overset{\overset{\displaystyle O}{\parallel}}{\underset{\underset{\displaystyle O}{\parallel}}{Fe}}^{VI}\text{–}O^{-} \qquad \text{Fe covalence, 6} \qquad (3\text{-}12)$$
$$(d^6 s^2) \qquad\qquad\qquad (d^5 sp^2); \, S{=}2/2;$$
$$14\text{-}e^{-} \text{ system}$$

$$Fe + 3 \qquad\qquad\qquad\qquad \text{Fe covalence, 6} \qquad (3\text{-}13)$$
$$(d^5 sp^2); \, S{=}2/2; \, 14\text{-}e^{-} \text{ system}$$

$$(\text{porphyrin})Fe^{II} + \cdot \ddot{O} \cdot \longrightarrow (\text{por})Fe^{IV}{=}O \qquad \text{Fe covalence, 4} \qquad (3\text{-}14)$$
$$(d^6 sp) \qquad\qquad\qquad (d^6 sp); \, S{=}2/2; \, 16\text{-}e^{-} \text{ system}$$
$$(\text{Fe covalence, 2}) \qquad\qquad (\text{with an axial imidazole it}$$
$$\text{becomes an } 18\text{-}e^{-} \text{ system})$$

$$Fe + 5\,CO \longrightarrow (O{=}\overset{..}{C}{:})Fe^{VIII}{=}(C{=}O)_4 \qquad \text{Fe covalence, 8} \qquad (3\text{-}15)$$
$$(d^5 sp^2); \, S{=}0; \, 18\text{-}e^{-} \text{ system}$$

$$Fe + 2 \cdot C_5H_5 \; (\cdot Cp) \longrightarrow (Cp)-Fe^{II}-(Cp) \qquad\qquad \text{Fe covalence, 2} \quad (3\text{-}16)$$

$$(d^6sp); \; S=0; \; 18\text{-}e^- \text{ system}$$

The valence-bond concept of chemical bonding[3] begins with the proposition that two neutral atoms combine to form covalent electron-pair bonds via a 1:1 contribution of unpaired valence electrons from the two atoms (e.g., $H\cdot + \cdot \ddot{\underset{..}{Cl}} : \longrightarrow H-\ddot{\underset{..}{Cl}} :$). Although the term *valence* is used in organic chemistry to indicate the number of covalent bonds for an atom in a molecule [for example, the carbon atoms in diamond ($C_n$), $CH_4$, $CCl_4$, $CO_2$, and benzene ($C_6H_6$) have a valence of 4], some chemists use valence to mean the oxidation state (oxidation number, ionic valence, or charge number) of atoms in molecules. On this basis, the chemical formula for $CrO_3$ can be written $Cr^{6+}(O^{2-})_3$ with oxidation numbers of six (VI) and two (II) assigned to Cr and O. However, the bonding in $CrO_3$ is more reasonably formulated as covalent with 3 $\sigma$ bonds and 3 $\pi$ bonds between a chromium atom ($d^4s^2 \to d^3sp^2$) and 3 oxygen atoms ($s^2p^4$), and is similar to that in $CO_2$ (C, $sp^3$ and O, $s^2p^4$). In $CrO_3$ the six covalent bonds mean the valence for chromium is 6 (as it is 4 for carbon in $CO_2$).

This confusion in nomenclature may be reduced by the use of *covalence* in place of valence and defined as the number of covalent bonds of an atom in a molecule. This term was first suggested by Langmuir to give a quantitative measure of the number of covalent bonds for an atom in a molecule.[10] Thus, for the molecules $FeCl_2$, $FeO$, $Fe(Cp)_2$, (porphyrin)Fe, and $Fe(Me)_2$ the covalence of iron is two; for $FeCl_3$, $Fe_2O_3$, $Fe(OH)_3$, and (porphyrin)Fe(OH) the covalence of iron is three; for (porphyrin)Fe(O) ("ferryl iron") the covalence of iron is four; for $FeO_4^{2-}$ (ferrate) and $Fe(catecholate)_3$ the covalence of iron is six; and for $Fe(CO)_5$ the covalence of iron is eight. In each of these compounds the iron atom is uncharged and has eight valence electrons ($3d^64s^2 \to d^6sp \to d^5sp^2$). For these examples, the traditionally used formal oxidation states of iron [II, III, IV, VI, and VIII (or 0), respectively] are the same as their covalences (number of covalent bonds). However, the iron in (porphyrin)$Fe^{(III)}(OH_2)^+$ ($d^5sp^2$) has a covalence of three, a formal oxidation state of three, and a charge of 1+ via the covalently bound $H_2O$. In the present discussion Roman numeral superscripts associated with the metals in the formulas for their compounds and complexes indicate their covalence (number of covalent bonds), *not* their oxidation state or number.

## Covalent bonding in oxygen containing molecules

With simple organic compounds covalent bonds result from the combination of the four unpaired electrons in the $sp^3$-hybridized orbitals of carbon with the unpaired electrons of hydrogen or oxygen.

$$\cdot \ddot{\underset{\cdot}{C}} \cdot + 4\,H\cdot \longrightarrow H - \underset{\underset{H}{|}}{\overset{\overset{H}{|}}{C}} - H \ (4\ \sigma\ \text{bonds}) \tag{3-17}$$

$$\cdot \ddot{\underset{\cdot}{C}} \cdot + 2 \cdot O\cdot \longrightarrow O{=}C{=}O \ (2\ \sigma\ \text{bonds and}\ 2\ \pi\ \text{bonds}) \tag{3-18}$$

Likewise, atomic hydrogen forms covalent bonds with atomic oxygen (without charge transfer)

$$2\,H\cdot + \cdot O\cdot \longrightarrow H - O - H \tag{3-19}$$

$$H\cdot + \cdot O\cdot \longrightarrow H - O\cdot \tag{3-20}$$

Addition of an electron to HOH or HO· goes to the atom with the greatest electron affinity (Table 3-1; O, 1.46 eV and H, 0.75 eV)

$$HOH + e^- \longrightarrow HO^- + H\cdot, \quad E°, \text{-2.9 V versus NHE} \tag{3-21}$$

$$HO\cdot + e^- \longrightarrow HO^-, \quad E°, \text{+1.9 V versus NHE} \tag{3-22}$$

The covalent bonding of atomic oxygen (·O·, 2 $s^2p^4$) with carbon and hydrogen is representative of its bonding to all elements. In neutral molecules without an oxygen-centered radical (HO· and MeO·) the covalence of oxygen is two (e.g., H–O–H, H–O–O–H, ·N=O, H–O–Cl, O=C=O, Me–O–H). In contrast, the covalence of oxygen in the hydronium ion ($H_3O^+$) is three, and in hydroxide ion (HO$^-$) is one. The valence-bond approach can be extended to molecular oxygen (dioxygen, ·O₂·); [·$\ddot{\underset{\cdot\cdot}{O}}$–$\ddot{\underset{\cdot\cdot}{O}}$· ↔ ·$\ddot{O}$≡$\underset{\cdot\cdot}{O}$·]), which has a bond order of two; nitric acid

(H–O–N$\overset{\displaystyle O}{\underset{\displaystyle O}{\Big\|}}$ ); pyridine N-oxide (⬡ N=O); and carbon dioxide (O=C=O).

The C–O bond energy ($\Delta H_{DBE}$) in $CO_2$ is 130 kcal, and the N–N bond energy in N≡N is 227 kcal. Hence, the bond energy for CO (isoelectronic with $N_2$), 257 kcal, prompts the formulation of an eight-electron covalent bond [C≡O: or two 3-electron bonds plus a σ bond].

The bond orders for ground-state molecular oxygen (dioxygen, $^3O_2$), for superoxide ion ($O_2^{-\cdot}$), and for (HOOH) are 2, 1.5, and 1, respectively. Table 3-3 provides molecular-orbital diagrams for dioxygen ($^3\Sigma_g^-$ , $^1\Delta_g$, and $^1\Sigma_g^+$ ) and superoxide ion, which illustrate the bi-radical character of $^3O_2(\cdot O_2\cdot)$ and the delocalized spin density of $O_2^{-\cdot}$ (one-half spin per oxygen).

The bonding of an oxygen atom in a water molecule (H–$\ddot{O}$–H) involves an octet of electrons about the oxygen atom (closed valence-electron shell) with two covalent sigma (σ) H–O bonds ($1s$-$2p_{x,y}$) at a dihedral angle of 104° [~90° for the $p_{x,y}$ orbital geometry of ground-state atomic oxygen (·O·)]. This bivalent bond

Table 3-3  Molecular-Orbital Diagrams for $O_2$ and $O_2^-$.

| Orbital | O₂ $^3\Sigma_g^-$ | $^1\Delta_g$ | $^1\Sigma_g^+$ | $O_2^-$ |
|---|---|---|---|---|
| $2p\sigma_u$ | — | — | — | — |
|  |  |  | ↑↓ | ↑ |
| $2p\pi_g$ | ↑ ↑ | ↑↓ |  | ↑↓ |
| $2p\pi_u$ | ↑↓ ↑↓ | ↑↓ ↑↓ | ↑↓ ↑↓ | ↑↓ ↑↓ |
| $2p\sigma_g$ | ↑↓ | ↑↓ | ↑↓ | ↑↓ |
| $2s\sigma_u$ | ↑↓ | ↑↓ | ↑↓ | ↑↓ |
| $2s\sigma_g$ | ↑↓ | ↑↓ | ↑↓ | ↑↓ |

order for oxygen in its compounds with nonmetals and metals is general, and is illustrated by examples in Tables 3-4 and 3-5. The latter presents the traditional ionic formulas for metal oxides and reformulates them with uncharged metal atoms  and  bivalent atomic oxygen atoms ($\cdot O \cdot$) and oxy anions [$\cdot O^-$, $Mn^{VII}(O)_3$ $(O^-)^-$]. The addition of covalently bonded $Na_2^I O(s)$ to water results in the transfer of an electron from each sodium atom to the oxygen of a water molecule and internally to the oxygen atom

$$Na_2^I O(s) + HOH \xrightarrow{\;H_2O\;} 2\,Na^+(aq) + 2\,HO^-(aq) \qquad (3\text{-}23)$$

Table 3-4 Examples of Oxygen Covalent Bonding with Nonmetals

| C (Si, Sn, Pb) | N (P, As, Sb) | S (Se, Te) | Cl (Br, I) |
|---|---|---|---|
| Me-O-H | H₂N-O-H | O=S=O | H-O-Cl |
| Me-O-Me | :N≡N=Ö: | | Cl-O⁻ |
| | ·Ṅ=O | | |
| Me₂C=O | O=N=O | | |
| | | | Cl-OO· |
| Me-C(=O)(H) | (O)₂N-N(O)₂ | | |
| Me-O-O-H | HO-N̈=O | | |
| C≡O: | HO-N(=O)(O) | | |
| O=C=O | | | |

Likewise the reaction of $Na_2^IO(s)$ with $CO_2(g)$ yields covalently bonded sodium carbonate, which dissociates heterolytically in aqueous solutions

$$Na_2^IO(s) + CO_2(g) \longrightarrow Na^I\text{-}O\text{-}\overset{O}{\underset{}{\overset{\parallel}{C}}}\text{-}O\text{-}Na^I(s) \xrightarrow{H_2O} Na^+(aq) + CO_3^{2-}(aq) \quad (3\text{-}24)$$

Metal–oxygen compounds can be directly synthesized by atom–atom reactions, especially when sources of atomic oxygen are utilized.

$$m\text{-ClPhC(O)OOH} \longrightarrow m\text{-ClPhC(O)OH} + [O] \quad (3\text{-}25)$$

$$HOOH \longrightarrow HOH + [O] \quad (3\text{-}26)$$

$$O_3 \longrightarrow O_2 + [O] \quad (3\text{-}27)$$

$$HOCl \longrightarrow HCl + [O] \quad (3\text{-}28)$$

Table 3-5  Reformulation of Metal–Oxygen Compounds and Complexes with Covalent Bonding

| Traditional formula | Covalent formula[a] | (valence-electron hybridization) |
|---|---|---|
| $(Li^+)_2(O^{2-})$ | $Li^I\text{-}O\text{-}Li^I$ | ($s$) |
| $(Ba^{2+})(O^{2-})$ | $Ba^{II}=O$ | ($sp$) |
| $(Al^{3+})_2(O^{2-})_3$ | $O=Al^{III}\text{-}O\text{-}Al^{III}=O$ | ($sp^2$) |
| $(Si^{4+})(O^{2-})_2$ | $O=Si^{IV}=O$ | ($sp^3$) |
| $(Pb^{2+})(O^{2-})$ | $Pb^{II}=O$ | ($sp$) |
| $(Pb^{4+})(O^{2-})_2$ | $O=Pb^{IV}=O$ | ($sp^3$) |
| $(Pb^{4+})(^-OAc)_4$ | $Pb^{IV}(\text{–}OAc)_4$ | ($sp^3$) |
| $(Ti^{4+})(O^{2-})_2$ | $O=Ti^{IV}=O$ | ($d^2sp$) |
| $(V^{5+})_2(O^{2-})_5$ | | ($d^3$sp) |
| $(V^{4+})(O^{2-})\,(^-OH)_2$ | ($S=1/2$) | ($d^3sp$) |
| $(Cr^{6+})(O^{2-})_3$ | | ($d^4sp$) |
| $[(Mn^{7+})(O^{2-})_4]^-$ | $(O=)_3Mn^{VII}\text{-}O^-$ ($S=0$) | ($d^5sp$) |
| $(Mn^{4+})(O^{2-})_2$ | $O=Mn^{IV}=O$ ($S=3/2$) | ($d^5sp$) |
| $(Mn^{2+})(O^{2-})$ | $Mn^{II}=O$ ($S=5/2$) | ($d^5sp$) |

Table 3-5 (cont.)

| | | |
|---|---|---|
| $Fe^{6+})(O^{2-})_4]^{2-}$ | $^-O-\overset{\overset{O}{\|}}{\underset{\underset{O}{\|}}{Fe}}^{VI}-O^-$ $(S = 2/2)$ | $(d^5sp^2)$ |
| $(Fe^{3+})_2(O^{2-})_3$ | $O=Fe^{III}-O-Fe^{III}=O$ $(S = 5/2)$ | $(d^5sp^2)$ |
| $(Fe^{2+})(O^{2-})$ | $Fe^{II}=O$ $(S = 4/2)$ | $(d^6sp)$ |
| $(Fe^{5+})(TPP^{2-})(O^{2-})$ Compound I, HRP | $(TPP^{+}\cdot)\,Fe^{IV}=O$ $(S = 3/2)$ | $(d^6sp)$ |
| $(Fe^{4+})(TPP^{2-})(O^{2-})$ Compound II, HRP | $(TPP)Fe^{IV}=O$ $(S = 2/2)$ | $(d^6sp)$ |
| $(Fe^{5+})(TPP^{2-})(O^{2-})(RS^-)$ Compound I, cyt $P$-450 | $(TPP)(RS)Fe^V=O$ $(S = 1/2)$ | $(d^5sp^2)$ |
| $(Co^{3+})_2(O^{2-})_3$ | $O=Co^{III}-O-Co^{III}=O$ $(S=0)$ | $(d^6sp^2)$ |
| $(\cdot Co^{2+})(O^{2-})$ | $\cdot Co^{II}=O$ | $(d^7sp)$ |
| $(Au^+)_2(O^{2-})$ | $Au^L-O-Au^I$ | $(d^{10}s)$ |
| $(Au^{3+})_2(O^{2-})_3$ | $O=Au^{III}-O-Au^{III}=O$ | $(d^8sp^2)$ |
| $(Pt^{2+})(O^{2-})$ | $Pt^{II}O$ | $(d^8sp)$ |
| $(Pt^{4+})(O^{2-})_2$ | $O=Pt^{IV}=O$ | $(d^8sp)$ |
| $(Mo^{5+})_2(O^{2-})_5$ | $\overset{O}{\underset{O}{\diagup}}\!\!\!\diagdown\overset{\cdot\,V}{Mo}-O-\overset{\cdot\,V}{Mo}\diagup\!\!\!\underset{O}{\diagdown}\overset{O}{}$ $(S = 1/2)$ | $(d^4sp)$ |

---

[a] Superscript Roman numerals represent *covalence* (number of covalent bonds, not oxidation state or number).

$$HOIO_3 \longrightarrow HOIO_2 + [O] \tag{3-29}$$

The dissolution and oxidation of the elemental manganese in steel alloys to permanganate [$HOMn^{VII}O_3$] by $HOIO_3$ illustrates such oxygen-atom-transfer chemistry and the associated increase in the covalence of Mn.

$$Mn(s) + [O] \longrightarrow Mn^{II}{=}O(s) \xrightarrow{[O]} O{=}Mn^{IV}{=}O(s) \tag{3-30}$$

$$2\,Mn^{IV}(O)_2(s) + 3\,[O] \longrightarrow (O)_3Mn^{VII}OMn^{VII}(O)_3 \xrightarrow{H_2O}$$
$$2\,H_3O^+(aq) + 2\,(O)_3Mn^{VII}O^-(aq) \tag{3-31}$$

## Covalent bond energies for oxygen in molecules

Although the gas-phase dissociative $X$–O bond energies ($\Delta H_{DBE}$) are known for many diatomic and nonmetallic molecules (see Table 3-6 for illustrative examples),[9] the metal–oxygen bond energetics for transition-metal complexes, organometallics, and metalloproteins in the condensed phase are not

Table 3-6  Gas-Phase Dissociative Bond Energies ($\Delta H_{DBE}$) for Oxygen-Containing Molecules[a]

| Bond | $\Delta H_{DBE}$ (kcal mol$^{-1}$) | Bond | $\Delta H_{DBE}$ (kcal mol$^{-1}$) |
|---|---|---|---|
| O=O | 119 | H–OH | 119 |
| HO–OH | 51 | H–O· | 102 |
| $t$-BuO–OBu-$t$ | 52 | H–OOH | 89 |
| PhCH$_2$O–Me | 67 | | |
| OC=O | 127 | H–OO· | 47 |
| C≡O | 257 | H–O$^-$ | 115 |
| Ba$^{II}$=O | 134 | H–OO$^-$ | 63 |
| Mn$^{II}$=O | 96 | | |
| $^+$Mn$^{II}$=O | 57 | H–OMe | 104 |
| Fe$^{II}$=O | 93 | H–OAc | 106 |
| $^+$Fe$^{II}$=O | 71 | H–Bu-$t$ | 105 |
| $^+$Fe$^L$–OH | 73 | | |
| Fe$^L$–Cl | 84 | | |
| (CO)$_4$Fe$^{VIII}$–CO | 41 | | |
| ·Co$^{II}$=O | 92 | | |
| Ni$^{II}$=O | 91 | | |

[a] Ref. 9.

understood. With the latter systems the traditional thermochemical methods of evaluation are tedious and imprecise.

The need for quantitative bond-energy data for oxygen-containing molecules in the condensed phase has prompted the development of an evaluation procedure that is based on electron-transfer thermodynamics. The approach is illustrated for H–O bonds and for Fe–O bonds, but is applicable to any X–O bond for which appropriate electron-transfer potentials are available. Table 3-7a summarizes the one-electron standard reduction potentials for $H_3O^+$, HO·, HOH, $O_2$, HOO·, $O_2^-$, ·O·, and O·· in aqueous solutions (see Chapter 2).[11–14]

Combination of the appropriate electron-transfer half-reactions (Table 3-7a) provides an expression for the free energy of bond formation [$-\Delta G_{BF} = nF (\Delta E°)_{reac}$]. For example, in the case of the H–OH bond, the energetics for the reduction of a proton in the absence and presence of HO· are given by

$$H_3O^+ + e^- \longrightarrow H· + H_2O, \quad E°, -2.10 \text{ V versus NHE} \tag{3-32}$$

and

$$HO· + H_3O^+ + e^- \longrightarrow H–OH + H_2O, \quad E°, +2.72 \text{ V} \tag{3-33}$$

Subtraction of Eq. 3-32 from Eq. 3-33 gives

$$HO· + H· \longrightarrow H–OH, \quad (\Delta E°)_{reac}, +4.82 \text{ V} \tag{3-34}$$

which provides a measure of the H–OH bond energy.

$$-\Delta G_{BF} = nF (\Delta E°)_{reac} = [(4.82)] [23.1 \text{ kcal (eV)}^{-1}] = 111 \text{ kcal} \tag{3-35}$$

Analogous arguments are used to evaluate the free energies of bond formation ($-\Delta G_{BF}$) for the H–O·, H–O$^-$, H–OOH, H–OO$^-$, and H–OO· bonds, which are summarized in Table 3-7b.

An equivalent approach has been used to evaluate metal–ligand covalent bond-formation free energies ($-\Delta G_{BF}$) for metal complexes $(ML_n)$[15] and metal-oxygen compounds.[16] Thus, for $ML_3$ complexes the differential redox potentials ($\Delta E°' = E°'_{L/L^-} - E°'_{ML_3/ML_3^-}$) of their ligand-centered electron-transfer processes provide a measure of the $L_2M–L$ bond energy [$-\Delta G_{BF} = (\Delta E°') 23.1 \text{ kcal mol}^{-1}$].[15] Table 3-8 illustrates such an approach for iron complexes and summarizes their apparent free energies of covalent bond formation ($M–L$, $-\Delta G_{BF}$).[11, 15–19] Additional metal–ligand covalent-bond-formation free energies ($-\Delta G_{BF}$) for manganese, iron, and cobalt complexes are summarized in Table 3-9.[15, 16, 20]

Although the uncharged tris 3,5-di-$t$-butyl-catecholate complex of iron [Fe(DTBC)$_3$] has been extensively studied,[17,21,22] the proposed bonding in these reports is unclear. The most common formulation is as an ionic salt between iron(3+) and three semiquinone anion radicals, $Fe^{3+}(DTBSQ^{-·})_3$. However, the magnetic moment is 2.9 BM (consistent with an $S = 2/2$ spin state)[17,22] and the

Table 3-7 Standard Aqueous Reduction Potentials for Protons, Water, and Oxygen Species; and the Free Energy of Bond Formation ($-\Delta G_{BF}$) for Several Hydrogen–Oxygen Species

a. Reduction potentials in $H_2O$ (standard states, unit molality)[a]

| Redox couple | $E°$, V versus NHE | Redox couple | $E°$, V versus NHE |
|---|---|---|---|
| $HO\cdot + H_3O^+ + e^- \rightarrow H\text{-}OH + H_2O$ | +2.72 | $\cdot O^- + HOH + e^- \rightarrow H\text{-}O\cdot + HO^-$ | +1.77 |
| $\cdot O\cdot + H_3O^+ + e^- \rightarrow H\text{-}O\cdot + H_2O$ | +2.14 | $O_2{\cdot}^- + HOH + e^- \rightarrow H\text{-}OO\cdot + HO^-$ | +0.20 |
| $HOO\cdot + H_3O^+ + e^- \rightarrow H\text{-}OOH + H_2O$ | +1.44 | $H_3O^+ + e^- \rightarrow H\cdot + H_2O$ | -2.10 |
| $\cdot O_2{}^- + H_3O^+ + e^- \rightarrow H\text{-}OO\cdot + H_2O +$ | +0.12 | $HOH + e^- \rightarrow H\cdot + HO^-$ | -2.93 |

b. Free energies of bond-formation ($-\Delta G_{BF}$)

| Bond | Redox evaluation $-\Delta G_{BF}$ (kcal mol$^{-1}$ eV$^{-1}$) | $-\Delta G_{BF}$ (kcal mol$^{-1}$) |
|---|---|---|
| H-OH | [2.72 - (-2.10)] 23.1 | 111 |
| H-O· | [2.14 - (-2.10)] 23.1 | 98 |
| H-O⁻ | [1.77 - (-2.93)] 23.1 | 109 |
| H-OOH | [1.44 - (-2.10)] 23.1 | 82 |
| H-OO· | [0.20 - (-2.93)] 23.1 | 72 |
| H-OO⁻ | [0.12 - (-2.10)] 23.1 | 51 |

[a] Refs. 11-14.

Table 3-8   Redox Potentials for Iron Complexes and Ligands, and Their Apparent Metal–Ligand Covalent-Bond-Formation Free Energies ($-\Delta G_{BF}$)

| Complex $(FeL_n)^a$ | $E_{1/2}$, V versus NHE[b] | | Bond | $-\Delta G_{BF}$ (kcal mol$^{-1}$)[c] |
| --- | --- | --- | --- | --- |
| | $\cdot L/L^-$ | $FeL_n/FeL_n^-$ | | |
| **a. MeCN[d]** | | | | |
| $Fe^{III}Cl_3$ ($d^5sp^2$) | +2.11 | +0.34 | $Cl_2Fe-Cl$ | 41 |
| $Fe^{III}(8Q)_3{}^e$ | +0.21 | −0.41 | $(8Q)_2Fe-O$(of 8Q) | 14 |
| $Fe^{III}(acac)_3{}^e$ | +0.55 | −0.42 | $(acac)_2Fe-O$(of acac) | 18 |
| $Fe^{III}(PA)_3{}^e$ | +1.50 | +0.20 | $(PA)_2Fe-O$(of PA) | 30 |
| $Fe^{III}(DTBC)_2^{-f}$ | +0.12 | −0.24 | $^-(DTBC)Fe-O$(of DTBC) | 8 |
| $Fe^{III}(TDT)_2^{-g}$ | +0.19 | −0.59 | $^-(TDT)Fe-S$(of TDT) | 18 |
| $Fe^{III}(bpy)_3{}^{3+e}$ | +2.32(+/0) | +1.30 | $^{2+}(bpy)_2Fe-N$(of bpy$^+$) | 24 |
| $Fe^{III}(MeCN)_4{}^{3+\,e}$ | >+2.5 (+/0) | +1.27 | $^{2+}(Ph_3PO)_3Fe-(OPPh_3{}^+)$ | >28 |
| $Fe^{III}(OPPh_3)_4{}^{3+\,e}$ | >+2.5 (+/0) | +1.85 | $^{2+}(MeCN)_3Fe-(MeCN^+)$ | >15 |
| $(Cl_8TPP)Fe^{III}Cl^e$ | +2.11 | −0.02 | $(Cl_8TPP)Fe-Cl$ | 49 |
| $(Cl_8TPP)Fe^{III}(OH)^e$ | +0.92 ($\cdot OH/^-OH$) | −0.48 | $(Cl_8TPP)Fe-OH$ | 32 |
| $(Cl_8TPP)Fe^{IV}(O)^h$ | +0.67 ($\cdot O/O\cdot^-$) | −0.30 | $(Cl_8TPP)Fe{=}O \rightarrow$ | 22 (π bond) |
| $Fe^{VI}(DTBC)_3{}^f$ ($d^5sp^2$; $S=2/2$) | +0.12 | −0.29 | $(Cl_8TPP)Fe-O-$ $(DTBC)_2Fe-O$(of DTBC) | 9 |

b. Aqueous, pH 14[d,i]

| | | | | |
|---|---|---|---|---|
| $Fe^{VI}(O)_2(O\cdot)_2^{2-}$; $S=2/2$ (d⁵sp²) | +1.43 ($\cdot O/O\cdot$) | +0.55 | $(^-O)_2(O)Fe{=}O$ | 20 ($\pi$ bond) |
| $Fe^{III}(O)(O\cdot)^-$ (d⁵sp²) | +1.77 ($O\cdot^-/2\text{-}OH$) | −0.7 | $(O)Fe{-}O\cdot$ | 57 |
| $Fe^{II}(OH)(O\cdot)^-$ (d⁶sp) | +1.89 ($\cdot OH/^-OH$) | +0.8 | $(^-O)Fe{-}OH$ | 62 |
| $Fe^{III}(CN)_6^{3-}$ (d⁵sp²) | +2.4 | +0.36 | $^{3-}(CN)_5Fe{-}CN$ | 47 |

c. Aqueous, pH 0 [d,i]

| | | | | |
|---|---|---|---|---|
| $^{2+}(H_2O)_2Fe^{III}(OH_2)^+$ (d⁵sp²) | +2.72 ($\cdot OH,H^+/H_2O$) | +0.77 | $^{2+}(H_2O)_2Fe{-}OH_2^+$ | 45 |

[a] Key: 8Q, 8-quinolinate; acac, acetylacetonate; PA, picolinate; DTBC, 3,5-di-*t*-butyl-catecholate; TDT, 3,4-toluene-dithiolate; bpy, 2,2-bipyridine.
[b] $E_{NHE} = E_{SCE} + 0.24$ V.
[c] Ref. 15.
[d] The superscript Roman numerals indicate the covalence (number of covalent bonds) for the metal; *not* the oxidation state.
[e] Ref. 16.
[f] Ref. 17.
[g] Ref. 18.
[h] Ref. 19.
[i] Ref. 11.

Table 3-9 Apparent Metal–Ligand Covalent-Bond-Formation Free Energies (-$\Delta G_{BF}$) for Several Manganese, Iron, and Cobalt Complexes[a,b]

| Mn–L bond | -$\Delta G_{BF}$ (kcal mol$^{-1}$) | Fe–L bond | -$\Delta G_{BF}$ (kcal mol$^{-1}$) | Co–L bond | -$\Delta G_{BF}$ (kcal mol$^{-1}$) |
|---|---|---|---|---|---|
| (8Q)$_2$Mn$^{III}$–8Q | 6 | (8Q)Fe$^{III}$–8Q | 15 | (8Q)$_2$Co$^{III}$–8Q | 16 |
| (acac)$_2$Mn$^{III}$–acac | 9 | (acac)$_2$Fe$^{III}$–acac | 23 | (acac)$_2$Co$^{III}$–acac | 21 |
| (PA)$_2$Mn$^{III}$–PA | 22 | (PA)$_2$Fe$^{III}$–PA | 31 | (PA)$_2$Co$^{III}$–PA | 35 |
| (Cl$_8$TPP)Mn$^{III}$–OH | 25 | (Cl$_8$TPP)Fe$^{III}$–OH | 31 | (Cl$_8$TPP)Co$^{III}$–OH | 25 |
| (Cl$_8$TPP)Mn$^{III}$–OO- | 17 | (Cl$_8$TPP)Fe$^{III}$–OO- | 12 | (Cl$_8$TPP)Co$^{III}$–OO- | 7 |
| (Cl$_8$TPP)Mn$^{IV}$=O | 87 | (Cl$_8$TPP)Fe$^{IV}$=O | 67 | | |
| (Cl$_8$TPP+)Mn$^{IV}$=O | 35 | (Cl$_8$TPP+)Fe$^{IV}$=O | 46 | | |
| | | (TPP)Fe$^{II}$–OFe$^{III}$(TPP) | 45 | | |
| [(bpy)$_2$Mn$^{III}$–bpy]$^{3+}$ | 23 | [(bpy)$_2$Fe$^{III}$–bpy]$^{3+}$ | 29 | [(bpy)$_2$Co$^{III}$–bpy]$^{3+}$ | 46 |

[a]See footnote of Table 3-8 for key to abbreviations.
[b]Refs. 15–20.

electrochemistry indicates a ligand-centered reduction.[17] Both of these characteristics are analogous to ferrate dianion $Fe^{VI}(O)_4^{2-}$ ($d^5sp^2$; $S=2/2$; covalence, 6). Thus, the iron center ($d^5sp^2$, eight unpaired electrons) forms six covalent Fe–O bonds with three catechols (equivalent to the six H–O bonds in three $DTBCH_2$ molecules), which leaves two unpaired electrons at the iron center ($S = 2/2$) (Table 3-8).

The results from the estimation of covalent bond energies from ligand-centered redox potentials for several iron and manganese compounds are compared to gas-phase values in Table 3-10.[9,19,23,24] The self-consistency is impressive, and indicates that the iron–ligand bond energy is reduced (1) by

Table 3-10 Comparison of Gas-Phase and Solution-Phase Bond-Formation Free Energies ($-\Delta G_{BF}$) for Iron and Manganese Compounds[a]

| Gas phase | $-\Delta G_{BF}$ (kcal mol$^{-1}$)[b] | Solution phase | $-\Delta G_{BF}$, (kcal mol$^{-1}$) |
|---|---|---|---|
| $Fe^{II}=O$ ($d^6sp$) | 85 | $(Cl_8TPP)Fe^{IV}=O$ ($d^6sp$) | 67 |
| $^+Fe^{II}=O$ ($d^5sp$) | 61 | $(Cl_8TPP\cdot)Fe^{IV}=O$ ($d^6sp$) | 46 |
| $^+Fe^{I}-OH$ ($d^6s$) | 65 | $(Cl_8TPP)Fe^{III}-OH$ ($d^5sp^2$) | 32 |
| | | $^{2+}(H_2O)_2Fe^{III}-OH$ ($d^5sp^2$) | 45 |
| $Mn^{II}=O$ ($d^5sp$) | 88 | $(Cl_8TPP)Mn^{IV}=O$ ($d^5sp$) | 87 |
| $^+Mn^{II}=O$ ($d^5s$) | 49 | $(Cl_8TPP^+)Mn^{V}=O$ ($d^5sp$) | 35 |
| $Fe^{I}-Cl$ ($d^6sp$) | 76 | $(Cl_8TPP)Fe^{III}-Cl$ ($d^5sp^2$) | 49 |
| | | $Cl_2Fe^{III}-Cl$ ($d^5sp^2$) | 41 |
| $(CO)_4Fe^{VIII}-(CO)$ | 33 | $(CO)_4Fe^{VIII}-(CO)$ | 34 |
| ($d^5sp^2$) | | $(Cl_8TPP)Fe^{IV}-(CO)$ ($d^6sp$) | 16 |
| $(Cp)Fe^{II}-Cp$ ($d^6sp$) | 83 | | |
| $V^{II}=O$ ($d^3sp$) | 142 | | |
| $^+V^{II}=O$ ($d^2sp$) | 123 | | |
| $Cr^{II}=O$ ($d^4sp$) | 95 | | |
| $^+Cr^{II}=O$ ($d^5$) | 77 | | |
| $Co^{II}=O$ ($d^7sp$) | 84 | | |
| $^+Co^{II}=O$ ($d^6sp$) | 56 | | |
| $^+Co^{I}-OH$ ($d^6s$) | 63 | | |

[a] The superscript Roman numerals indicate the covalence (number of covalent bonds) for the metal; *not* the oxidation state.

[b] Refs. 9 and 23.; $-\Delta G_{BF} = \Delta H_{DBE} - T\,\Delta S_{DBE} \approx \Delta H_{DBE} - 8$ (at 25°C).

[c] Refs. 19 and 24.

The formation of the same iron–oxygen covalent bonds from either (1) oxidized iron plus oxy anions via electron-transfer (redox) reactions or (2) radical–radical coupling reactions is summarized in Table 3-11. The valence-electron hybridization for the iron center is included as well as the spin state and estimated covalent bond-formation free energy ($\Delta G_{BF}$). A similar set of reactions and data for iron–porphyrin compounds is presented in Table 3-12. Section a emphasizes that, just as the combination of a proton with a hydroxide ion yields a covalent H–OH bond (Table 3-11), (1) the combination of protons and porphyrin dianion ($Por^{2-}$) yields covalent porphine ($H_2Por$), and (2) the addition of Lewis acids ($Zn^{2+}$ or $Fe^{2+}$) to porphine ($H_2Por$) oxidatively displaces protons to give covalent-bonded $Zn^{II}Por$ and $Fe^{II}Por$.

Although this discussion has focused on iron compounds (with an emphasis on the nature of iron–oxygen bonds), closely similar arguments and conclusions can be made for all transition-metal compounds. To the extent that a convincing case has been made that all neutral and negatively charged iron–oxygen compounds have covalent bonds with elemental iron ($d^6s^2 \rightarrow d^6sp \rightarrow d^5sp^2$ hybridization, and covalency 2, 3, 4, 6, or 8) and oxygen-centered redox chemistry, the same conclusions are in order for iron–sulfur compounds (including the iron–sulfur proteins).[25,26] However, the covalency of iron is limited to two or three in such compounds ($d^6sp$ and $d^5sp^2$ hybridization; $Fe^{II}(SR)_4^{2-}$ and $Fe^{III}(SR)_4^{-}$, $[(RS)_2Fe\text{-}(\mu\text{-}S)_2\text{-}Fe(SR)_2]^{3-,2-}$, and $[(RS)_4Fe_4S_4]^{3-,2-,-}$), and the compounds have negative redox potentials [-0.6 V versus NHE for $Fe^{III}(TDT)_2^{-}$].

Additional X–O dissociative bond energies ($\Delta H_{DBE}$; $XO \longrightarrow X + O$) for oxygen-containing molecules are summarized in Table 3-13.[9,11, 27–29] These provide a thermodynamic measure of the propensity for oxygen-atom transfer. For example the dissociative bond energies ($\Delta H_{DBE}$) of the oxygen atom in the compound-I model [$Cl_8TPP^{+\cdot})Fe^{IV}=O$] and in ethylene oxide ($H_2C\overset{\displaystyle \frown O}{\phantom{=}}CH_2$) are 54 and 84 kcal mol$^{-1}$, respectively (Tables 3-10 and 3-13). Hence, the transfer of an oxygen atom from compound I to ethylene is an exothermic process ($-\Delta H_{reac} = 30$ kcal mol$^{-1}$)

$$(Cl_8TPP^{+\cdot})Fe^{IV}=O + H_2C=CH_2 \longrightarrow H_2C\overset{\displaystyle \frown O}{\phantom{=}}CH_2 + (Cl_8TPP)Fe^+ \quad (3\text{-}36)$$

The oxygenation of saturated hydrocarbons by methane mono-oxygenases (binuclear nonheme iron proteins, MMO) via the two-electron reduction of dioxygen,[30–32]

$$CH_4 + O_2 + 2\,H^+ + 2\,e^- \xrightarrow{\text{MMO}} CH_3OH + H_2O \quad (3\text{-}37)$$

72

Oxygen Chemistry

Table 3-11 Bonding and Valence-Electron Hybridization for Iron–Oxygen Compounds

| Lewis acid–base electron transfer | Covalent product species (valence electrons) $S$=spin state, $-\Delta G_{BF}$, (kcal mol$^{-1}$) | | Radical–radical coupling |
|---|---|---|---|

$H^+ + HO^-$ $\xrightarrow{K_f, 10^{14}}$ H–OH (1s) $S=0$ **111** $\longleftarrow$ $H\cdot + \cdot OH$

$Fe^{3+} + HO^-$ $(d^5)$ $\longrightarrow$ $^{I}Fe^{2+}-OH$ $(d^5s)$ $S=5/2$ **45** $\longleftarrow$ $Fe^{2+} + \cdot OH$ $(d^6)$

$Fe^{3+} + 2\,HOH$ $(d^5)$ $\xrightarrow{H_3O^+}$ $^{I}Fe^{2+}-OH$ $(d^5s)$ $S=5/2$ **45** $\longleftarrow$ $Fe^{2+} + \cdot OH$

$Fe^{2+}-OH + HO^-$ $(d^5s)$ $\xrightarrow{HOH}$ $^{II}Fe^+=O$ $(d^5sp)$ $S=5/2$ **61** $\longleftarrow$ (HOH) $Fe^+ + 2\,\cdot OH$ $(d^5s^2)$

$Fe^{2+}-OH + HOH$ $\xrightarrow{H_3O^+}$ $^{II}Fe^+=O$ $(d^5s)$ $S=5/2$ **61** $\longleftarrow$ (HOH) $Fe^+ + 2\,\cdot OH$

$Fe^+=O + HO^-$ $(d^5sp)$ $\longrightarrow$ $^{III}Fe(=O)(OH)$ $(d^5sp^2)$ $S=5/2$ $\longleftarrow$ $Fe=O + \cdot OH$ $(d^5sp)$ $S=4/2$

$Fe^+=O + 2\,HOH$ $\xrightarrow{H_3O^+}$ $^{III}Fe(=O)(OH)$ $(d^5sp^2)$ $S=5/2$ $\longleftarrow$ $Fe=O + \cdot OH$

$Fe(=O)(OH) + HO^-$ $\xrightarrow{HOH}$ $^{III}Fe(=O)(O^-)$ $(d^5sp^2)$ $S=5/2$ $\longleftarrow$ $Fe=O + O^-$

$2\,Fe^+=O + 2\,HO^-$ $(d^5sp)$ $\xrightarrow{HOH}$ $^{III}Fe(=O)-O-(O=)Fe^{III}$ $(d^5sp^2)$ $\longleftarrow$ $2\,Fe=O + \cdot O^- (2\,\cdot OH)$ $(d^6sp)$

$Fe^{2+} + 2\,HO^-$ $(d^6)$ $\xrightarrow{HOH}$ $^{II}Fe=O$ $(d^6sp)$ $S=4/2$ **85** $\longleftarrow$ (HOH) $Fe + 2\,\cdot OH$ $(d^6s^2)$

$Fe=O + HO^-$ $\longrightarrow$ $^{II}Fe(-O^-)(OH)$ $(d^6sp)$ $S=4/2$ $\longleftarrow$ $Fe + \cdot OH + O^-$ $(d^6s^2)$

Table 3-11 (cont.)

$$\underset{(d^5sp^2)}{Fe} \overset{O}{\underset{O^-}{\diagdown}} \quad \begin{array}{c} +\,2\,HOO^- \\ +\,1/2\,HOOH \end{array} \xrightarrow[\;\;\overset{\textstyle\frown}{HOH}\;\;]{} \quad \overset{O}{\underset{O}{\underset{\|}{\overset{\|}{O^- -Fe^{VI} - O^-}}}} \quad \xleftarrow[\;\;\overset{\textstyle\frown}{HOH}\;\;]{} \quad \underset{(d^6sp)}{Fe{=}O + 2\,^\cdot OH + 2\,O^-}$$
$$(d^5sp^2)\;S{=}2/2$$

---

Table 3-12 Bonding, Electron-Transfer Reactions, and Valence-Electronics for Iron-Porphyrin

---

a.  Bonding and synthesis in metal porphyrins

$$2\,H^+ + 2\;\; \overset{\bigcap}{N_-} \rightarrow 2\;\; \underset{H}{\overset{\bigcap}{N}}$$

$$2\,H^+ + Por^{2-} \rightarrow (H\text{-})_2Por\;[H_2Por]$$
$$(1s)$$

$$\underset{(d^{10})}{H_2Por} + \underset{}{Zn^{2+}} \rightarrow \underset{(d^{10}sp)}{{}^{II}Zn{=}Por\;[ZnPor]} + 2\,H^+$$

$$\underset{(d^6)}{H_2Por} + Fe^{2+} \rightarrow \underset{(d^6sp)}{{}^{II}FePor} + 2\,H^+$$

$$\underset{(d^5)}{H_2Por} + Fe^{3+} \rightarrow \underset{(d^5sp)}{{}^{II}Fe^+Por} + 2\,H^+$$

b.  Redox thermodynamics in acetonitrile

| | $E^{o\prime}$, V versus NHE |
|---|---|
| 1.  Oxidizing agents | |
| $H^+ + e^- \rightarrow H\cdot$ | -1.58 |
| $(Cl_8TPP)^{II}Fe^+ + e^- \rightarrow (Cl_8TPP)Fe^{II}$ | +0.56 |
| $\quad\;\;(d^5sp)\qquad\qquad\qquad\;\;\;(d^6sp)$ | |
| 2.  Reducing agents | |
| $HO^- \rightarrow HO\cdot + e^-$ | +0.92 |
| $HOO^- \rightarrow HOO\cdot + e^-$ | -0.16 |
| $O_2^{-\cdot} \rightarrow \cdot O_2\cdot + e^-$ | -0.66 |
| $O_2^{2-} \rightarrow O_2^{-\cdot} + e^-$ | <-2.8 |
| $O^{-\cdot} \rightarrow \cdot O\cdot + e^-$ | +0.46 |

Table 3-12 (cont.)

$$Cl^- \rightarrow Cl\cdot + e^-$$                              +2.24
$$PhCH_2S^- \rightarrow PhCH_2S\cdot + e^-$$                     +0.24
$$Bu^- \rightarrow Bu\cdot + e^-$$                              -2.8

c. Covalent-bond formation and valence-electron hybridization

| Lewis acid–base electron transfer | Product species (valence electrons), $S$= spin state $-\Delta G_{BF}$, (kcal mol$^{-1}$) | Radical–radical coupling |
|---|---|---|
| $H^+ + HO^-$ ⟶ | H–OH $(1s)$ $S=0$ **111** | ⟵ $H\cdot + \cdot OH$ |
| $PorFe^+ + HO^-$ $(d^5sp)$ ⟶ | $PorFe^{III}$–OH $(d^5sp^2)$ $S=5/2$ **31** | ⟵ $PorFe + \cdot OH$ $(d^6sp)$ |
| $H^+ + O_2^{2-}$ ⟶ | H–OO$^-$ $(1s)$ $S=0$ **72** | ⟵ $H\cdot + O_2^{-}\cdot$ |
| $PorFe^+ + O_2^{2-}$ $(d^5sp)$ ⟶ | $PorFe^{III}$–OO$^-$ $(d^5sp^2)$ $S=5/2$ **12** | ⟵ $PorFe + O_2^{-}\cdot$ $(d^6sp)$ |
| $H^+ + O_2^{-}\cdot$ ⟶ | H–OO$\cdot$ $(1s)$ $S=1/2$ **51** | ⟵ $H\cdot + \cdot O_2\cdot$ |
| $PorFe^+ + O_2^{-}\cdot$ $(d^5sp)$ ⟶ | $PorFe^{IV}(O_2)$ $(d^6sp)$ $S=0$ | ⟵ $PorFe + \cdot O_2\cdot$ $(d^6sp)$ |
| $H^+ + HOO^-$ ⟶ | H–OOH $(1s)$ $S=0$ **81** | ⟵ $H\cdot + \cdot OOH$ |
| $PorFe^+ + HOO^-$ $(d^5sp)$ ⟶ | $PorFe^{III}$–OOH $(d^5sp^2)$ $S=5/2$ **8** | ⟵ $PorFe + \cdot OOH$ $(d^6sp)$ |
| $H^+ + O^{-}\cdot$ ⟶ | H–O$\cdot$ $(1s)$ $S=1/2$ **94** | ⟵ $H\cdot + \cdot O\cdot$ |
| $PorFe^+ + O^{-}\cdot$ ⟶ | $PorFe^{IV}=O$ $(d^6sp)$ $S=2/2$ **67** | ⟵ $PorFe + \cdot O\cdot$ |
| $H^+ + Cl^-$ ⟶ | H–Cl $(1s)$ $S=0$ **95** | ⟵ $H\cdot + \cdot Cl$ |

Table 3-12 (cont.)

| | | |
|---|---|---|
| PorFe + Cl⁻ ($d^5sp$) | ⟶ PorFe$^{III}$–Cl ($d^5sp^2$) $S=5/2$ **49** | ⟵ PorFe + ·Cl ($d^6sp$) |
| H⁺ + PhCH₂S⁻ | ⟶ H–SCH₂Ph (1$s$) $S=0$ **79** | ⟵ H· + PhCH₂S· |
| PorFe⁺ + PhCH₂S⁻ ($d^5sp$) | ⟶ PorFe$^{III}$–SCH₂Ph ($d^5sp^2$) $S=5/2$ **22** | ⟵ PorFe + PhCH₂S· ($d^6sp$) |
| H⁺ + Bu⁻ | ⟶ H–Bu (1$s$) $S=0$ **91** | ⟵ H· + Bu· |
| PorFe⁺ + Bu⁻ $d^5sp$ | ⟶ PorFe$^{III}$–Bu ($d^5sp^2$) $S=5/2$ **10** | ⟵ PorFe + Bu· ($d^6sp$) |

requires that a C–H bond be broken and replaced with a C–OH bond. Thus, a reaction sequence with an initial reduction of $O_2$

$$2\,H^+ + 2\,e^- + O_2 \longrightarrow HOOH \tag{3-38}$$

followed by a Fenton reaction would generate an HO· radical (see Chapters 4 and 5)

$$HOOH + LFe^{II} \longrightarrow LFe^{III}\text{–}OH + HO\cdot \tag{3-39}$$

These suggested product species are able to oxygenate methane via the reaction sequence

$$CH_4 + HO\cdot + LFe^{III}\text{–}OH \longrightarrow HOH + [\cdot CH_3 + LFe^{III}\text{–}OH]$$
$$\longrightarrow CH_3OH + LFe^{II} \tag{3-40}$$

Table 3-14 summarizes the dissociative bond energies ($\Delta H_{DBE}$) for H–C bonds of hydrocarbons and aldehydes, and H–X bonds in thiols and phenols. The bond-formation free energies ($-\Delta G_{BF}$) for H–OY bonds also are listed. Reference to these quantities confirms that HO· is the only oxy radical with a sufficiently large bond-formation energy to break the C–H bond of methane and thereby accomplish the net chemistry of the methane monooxygenases [Eq. (3-37)] via an (HO·)-generation pathway [Eqs. (3-38), (3-39), and (3-40)]. For substrates with weaker C–H bonds, other oxy radicals (YO·) have sufficient H–OY bond-formation energies to give net reactions,[33]

Table 3-13  Dissociative Bond Energies (XO $\longrightarrow$ X + O; $\Delta H_{DBE}$) for Oxygen-Containing Molecules

| Compound | $\Delta H_{DBE}$ (kcal mol$^{-1}$)[a] |
|---|---|
| $O_2 \longrightarrow 2\,O$ | 119 |
| $O_3 \longrightarrow O_2 + O$ | 25 |
| 1,2-Ph[C(O)OH][C(O)OOH] $\longrightarrow$ 1,2-Ph[C(O)OH]$_2$ + O | 31 |
| $BrO_4^- \longrightarrow BrO_3^- + O$ | 35 |
| $N_2O \longrightarrow N_2 + O$ | 39 |
| $HOOH \longrightarrow HOH + O$ | 39 |
| $ClO^- \longrightarrow Cl^- + O$ | 41 |
| t-BuOOH $\longrightarrow$ t-BuOH + O | 44 |
| $(HO)_5IO \longrightarrow HOIO_2 + 2\,H_2O + O$ | 46 |
| $BrO^- \longrightarrow Br^- + O$ | 47 |
| $ClO_2 \longrightarrow ClO^- + O$ | 51 |
| $HOCl \longrightarrow HCl + O$ | 52 |
| $IO^- \longrightarrow I^- + O$ | 61 |
| $ClO_4^- \longrightarrow ClO_3^- + O$ | 65 |
| $ClO_3^- \longrightarrow ClO_2^- + O$ | 68 |
| $MeONO_2 \longrightarrow MeONO + O$ | 72 |
| $HOSe(O)_2O^- \longrightarrow HOSe(O)O^- + O$ | 73 |
| $NO_3^- \longrightarrow NO_2^- + O$ | 83 |
| $\overset{\displaystyle \lceil O \rceil}{H_2C{-}CH_2} \longrightarrow H_2C{=}CH_2 + O$ | 84 |

Table 3-13 (cont.)

| | |
|---|---|
| $Me_2SO \longrightarrow Me_2S + O$ | 86 |
| $MeOH \longrightarrow Me\text{-}H + O$ | 89 |
| $Ph_3AsO \longrightarrow Ph_3As + O$ | 94 |
| $Mo^{VI}(O)_2(S_2CNEt_2)_2 \longrightarrow Mo^{IV}(O)(S_2CNEt_2)_2 + O$ | 94 |
| $(MeO)_2SO_2 \longrightarrow (MeO)_2SO + O$ | 108 |
| $Me_2SO_2 \longrightarrow Me_2SO + O$ | 111 |
| $HOH \longrightarrow H_2 + O$ | 117 |
| $^{2+}V^{II}O \longrightarrow V^{2+} + O$ | 117 |
| $HOS(O)_2O^- \longrightarrow HOS(O)O^- + O$ | 118 |
| $HOCN \longrightarrow HCN + O$ | 120 |
| $HOC(O)O^- \longrightarrow HC(O)O^- + O$ | 121 |
| $SO_4{}^{2-} \longrightarrow SO_3{}^{2-} + O$ | 123 |
| $MeC(O)OH \longrightarrow MeCH(O) + O$ | 123 |
| $Ph_3PO \longrightarrow Ph_3P + O$ | 125 |
| $CO_2 \longrightarrow CO + O$ | 127 |
| $^{2+}Ti^{II}O \longrightarrow Ti^{2+} + O$ | 127 |
| $NCO^- \longrightarrow NC^- + O$ | 127 |
| $n\text{-}Bu_3PO \longrightarrow n\text{-}Bu_3P + O$ | 139 |

---

[a] Refs. 9, 11, 27-19. $\Delta H_{DBE} = [E°O, 2\,H^+/H_2O\,\text{-}E°XO, 2\,H^+/X, H_2O]\,2 \times 23.1 + T\,\Delta S_{DBE} = 46.2(2.43 - 1.85)BrO_4^- + 8 = 35\,(BrO_4^- \longrightarrow BrO_3^- + O)$.

$$HOO\cdot + c\text{-}1,4\text{-}C_6H_8 \longrightarrow (c\text{-}1,3\ C_6H_7\ \cdot) + HOOH \qquad (3\text{-}41)$$

In summary, the bond energies for atomic oxygen ($\cdot O\cdot$) and dioxygen ($\cdot O_2\cdot$) with transition metals are much smaller than with either carbon or hydrogen (Tables 3-6, 3-7, 3-8, 3-9, and 3-14), and thereby provide the means to stabilize active forms of oxygen [$\cdot O\cdot \rightarrow MO$; $\cdot O_2\cdot \rightarrow M(O_2)$; $\cdot O_2\cdot \rightarrow MOOM$]. Although the weak metal–oxygen bonds attenuate the reactivity of atomic oxygen, this is more than compensated by its increased lifetime in condensed-phase reactors [the reaction of Eq. (3-36) is a good example]. In some cases the stabilization by transition metals can enhance the reactivity via a change in the spin state of atomic oxygen ($\cdot O \cdot \rightarrow MO \cdot$; see Chapter 4);

$$\cdot Co^{II}(bpy)_2^{2+} + HOOH \xrightarrow{\phantom{xxx}} [(bpy)_2^{2+}Co^{III}\text{-}O\cdot] \xrightarrow{H\cdot} (bpy)_2^{2+}Co^{III}O\text{-}H \qquad (3\text{-}42)$$
$$\searrow H_2O \qquad\qquad (-\Delta G_{BF}, 105 \text{ kcal mol}^{-1})$$

This appears to be the approach that has been utilized by Nature for the activation of $\cdot O_2 \cdot$ and HOOH for the dehydrogenation, mono-oxygenation, and dioxygenation of organic substrates. Consideration of the realm of oxygen-bond energetics provides the basis for the design of efficient, selective catalysts for oxygenation reactions. Chapters 4 and 6 discuss such transition-metal-induced activation of HOOH and $\cdot O_2 \cdot$, respectively.

Table 3-14  Dissociative Bond Energies ($\Delta H_{DBE}$) for H–X Molecules and Bond-Formation Free Energies (-$\Delta G_{BF}$) for H–OY Bonds[a]

| Bond | ($\Delta H_{DBE}$)(g), (kcal mol$^{-1}$) | Bond ($\cdot$OY) | (-$\Delta G_{BF}$)(aq), (kcal mol$^{-1}$) |
|---|---|---|---|
| H–CH$_3$ | 105 | H–OH ($\cdot$OH) | 111 |
| H–$n$-C$_3$H$_7$ | 100 | H–O$\cdot$ ($\cdot$O$\cdot$) | 98 |
| H–$c$-C$_6$H$_{11}$ | 95 | H–O$^-$ ($\cdot$O$^-$) | 109 |
| H–$t$-C$_4$H$_9$ | 93 | H–OOH ($\cdot$OOH) | 82 |
| H–CH$_2$Ph | 88 | H–OO$^-$ ($\cdot$OO$^-$) | 72 |
| H–$c$-C$_6$H$_7$ (CHD) | 73 | H–OO$\cdot$ ($\cdot$OO$\cdot$) | 51 |
| H–C(O)Ph | 87 | H–OMe ($\cdot$OMe) | 96 |
| | | H–OBu-$t$ ($\cdot$OBu-$t$) | 97 |
| H–Ph | 111 | H–OPh ($\cdot$OPh) | 79 |
| H–SH | 91 | H–OOMe ($\cdot$OOMe) | 82 |
| H–SCH$_3$ | 89 | H–OOBu-$t$ ($\cdot$OOBu-$t$) | 83 |
| H–SPh | 83 | H–OC(O)Me [$\cdot$OC(O)Me] | 98 |
| H–OPh | 86 | | |

[a] Ref. 9 and Table 3-7.

**The case against $O_2^+$·**

Although textbooks describe $O_2^+$· as a firmly established ionic form of dioxygen,[34] this is based on the characteristics of $O_2MF_6$ ($M$ = P, As, Sb, or Pt)[35–37] and the *assumption* that these compounds are salts $[(O_2^+ \cdot)MF_6^-]$. Such an ionic formulation is unsupported by their chemical and physical properties, and is inconsistent with the ionization potentials, electron affinities, and electronegativities of the constituent atoms (Table 3-1). Consideration of the physical data for these compounds ($O_2MF_6$) in combination with the bonding data for the pairs (FOF, HOH) and (FOOF, HOOH), and for HO· and HOO· provides a basis for (1) an estimation of the bond energetics of the FO· and FOO· molecules and (2) the conclusion that the more accurate despiction of $O_2MF_6$ molecules is a covalent adduct $(\cdot OOF)M^VF_5$. Table 3-15 includes a comparison of the bond energetics and bond order for H–O and F–O bonding in several molecules.

Table 3-15 Comparison and Estimation of the Bond Energetics and Bond Order for H/O and F/O Molecules

| Bond (bond order) | $\Delta H_{DBE}$ (kcal mol$^{-1}$) | Bond (bond order) | $\Delta H_{DBE}$ (kcal mol$^{-1}$) |
|---|---|---|---|
| H–OH (1) | 119 | F–OF (1)[a] | 40–50 |
| H–OOH (1) | 90 | F–OOF (1)[b] | 30–35 |
| HO–OH (1) | 51 | FO=OF (2) | 115–118 |
| H–O· (1) | 103 | F–O· (1) | 40 (est.) |
| H–OO· (1) | 47 | F···OO· ($\frac{1}{2}$) | 12 (est.) |
| HO···O· ($1\frac{1}{2}$) [H-Ö≡Q· ↔ H-Q̈-Q̈·] (0.25)   (0.75) | 85–90 | FO···O· ($2\frac{1}{2}$) [F-Ö≡Q· ↔ F-Q̈-Q̈· ] (0.75)   (0.25) | 136 (est.) |
| (O···O)⁻ ($1\frac{1}{2}$) [·Ö≡Q: ↔ ·O-Q̈:] (0.25)   (0.75) | 90 | | |
| (·O=O·) (2) [·Ö≡Q ↔ ·Q̈-Q̈ ] (0.50)   (0.50) | 119 | | |

[a] Ref. 38; F–O bond length, 1.41 Å.

[b] Ref. 39; F–O bond length, 1.58 Å; O–O bond length, 1.22 Å.

The electronegativities for F, O, and P (Table 3-1) provide a basis to estimate the charge transfer in the FO·, FOO·, and $PF_5$ molecules (see Table 3-2). These are: $\cdot O^{0.15+}-F^{0.15-}$, $\cdot OO^{0.08+}-F^{0.08-}$, and $P^{0.3+}(F^{0.06-})_5$. Hence, the combination of FO· and FOO· with $PF_5$ should prompt a small redistribution of charge [Eqs. (3-43) and (3-44)]

$$\cdot O^{0.15+}-F^{0.15-} + P\cdot^{0.3+}(F^{0.06-})_5 \rightarrow (\cdot O^{0.15+}-F^{0.10-}:)P^{0.25+}(F^{0.06-})_5 \qquad (3\text{-}43)$$

$$\cdot OO^{0.08+}-F^{0.08-} + P^{0.3+}(F^{0.06-})_5 \rightarrow (\cdot OO^{0.08+}-F^{0.07-}:)\ P^{0.29+}(F^{0.06-})_5 \qquad (3\text{-}44)$$

### References

1. Mullay, J. J. *Am. Chem. Soc.* **1986**, *108*, 1770.
2. Jug, K.; Epiotis, N. D.; Buss, S. J. *Am. Chem. Soc.* **1986**, *108*, 3640.
3. Pauling, L. C. *The Nature of the Chemical Bond*, 3rd ed. Ithaca, N. Y.: Cornell Univ. Press, 1960.
4. Lide, D. R. (ed.). *CRC Handbook of Chemistry and Physics*, 71st ed. Boca Raton, FL: CRC, 1990, p. 9-6.
5. Lide, D. R. (ed.). *CRC Handbook of Chemistry and Physics*, 71st ed. Boca Raton, FL: CRC, 1990, pp. 10-180–181, 10-210–211.
6. Pearson, R. G. J. *Am. Chem. Soc.* **1988**, *110*, 7684.
7. Sanderson, R. T. *Inorg. Chem.* **1986**, *25*, 3518.
8. Sanderson, R. T. *Polar Covalence*. New York: Academic Press, 1983.
9. Lide, D. R. (ed.). *CRC Handbook of Chemistry and Physics*, 71st ed. Boca Raton, FL: CRC, 1990, pp. 9-86–98.
10. L. C. Pauling; private communications, October 26, 1987.
11. Bard, A. J.; Parsons, R.; Jordon, J. *Standard Potentials in Aqueous Solution*. New York: Marcel Dekker, 1985.
12. Parsons, R. *Handbook of Electrochemical Constants*. London: Buttersworths, 1959, pp. 69–73.
13. Fee, J. A.; Valentine, J. S. In *Superoxide and Superoxide Dismutases* (Michelson, A. M., McCord, J. M.; Fridovich, I., eds.). New York: Academic Press, 1977, pp. 19–60.
14. Schwarz, H. A.; Dodson, R. W. J. *Phys. Chem.* **1984**, *88*, 3643.
15. Richert, S. A.; Tsang, P. K. S.; Sawyer, D. T. *Inorg. Chem.* **1989**, *28*, 2471.
16. Sawyer, D. T. *Comments Inorg. Chem.* **1990**, *10*, 129.
17. Jones, S. E.; Leon, L. E.; Sawyer, D. T. *Inorg. Chem.* **1982**, *21*, 3692.
18. Sawyer, D. T.; Srivatsa, G. S.; Bodini, M. E.; Schaefer, W. P.; Wing, R. M. J. *Am. Chem. Soc.* **1986**, *108*, 936.
19. Sugimoto, H.; Tung, H.-C.; Sawyer, D. T. J. *Am. Chem. Soc.* **1988**, *110*, 2465.
20. Tsang, P. K. S.; Sawyer, D. T. *Inorg. Chem.* **1990**, *29*, 0000.
21. Sofen, S. R.; Ware, D. C.; Cooper, S. R.; Raymond, K. N. *Inorg. Chem.* **1979**, *18*, 234.
22. Boone, S. R.; Purser, G. H.; Chang, H.-R.; Lowery, M. D.; Hendrickson, D. N.; Pierpont, G. G. J. *Am. Chem. Soc.* **1989**, *111*, 2292.
23. Kang, H.; Beauchamp, J. L. J. *Am. Chem. Soc.* **1986**, *108*, 5663.
24. Tsang, P. K. S.; Cofré, P.; Sawyer, D. T. *Inorg. Chem.* **1987**, *26*, 3604.

25. Carney, M. J.; Papaefthymiou, G. C.; Frankel, R. B.; Holm, R. H. *Inorg. Chem.* **1989**, *28*, 1497.
26. Han, S.; Czernuszewicz, R. S.; Spiro, T. G. *J. Am. Chem. Soc.* **1989**, *111*, 3496.
27. Watt, G. D.; McDonald, J. W.; Newton, W. E. *J. Less-Common Met.* **1977**, *54*, 415.
28. *Selected Values of Chemical Thermodynamic Properties.* Washington, D.C.: National Bureau of Standards, 1968; Technical Note 270-3.
29. Pedley, J. B.; Naytor, R. D.; Kirby, S. P. *Thermochemical Data of Organic Compounds*, 2nd ed. New York: Chapman and Hall, 1986.
30. Dalton, H. *Adv. Appl. Microbiol.* **1980**, *26*, 71.
31. Ericson, A.; Hedman, B.; Hodgson, K. O.; Green J.; Dalton, H.; Bentsen, J. G.; Beer, S. J.; Lippard, S. J. *J. Am. Chem. Soc.* **1988**, *110*, 2330.
32. Fox, B. G.; Surerus, K. K.; Munck, E.; Lipscomb, J. D. *J. Biol. Chem.* **1988**, *263*, 10553.
33. Sawyer, D. T.; McDowell, M. S.; Yamaguchi, K. S. *Chem. Res. Toxicol.* **1988**, *1*, 97.
34. Cotton, F. A.; Wilkinson, G. *Advanced Inorganic Chemistry*, 5th ed. NewYork: Wiley-Interscience, 1988, p. 462.
35. Bartlett, N. In *Preparative Inorganic Reactions* (Jolly, W. L., ed.). New York: Interscience, 1965, Vol. 2, p. 301.
36. Bartlett, N.; Beaton, S. P. *J. Chem. Soc. Chem. Commun.* **1966**, 167.
37. Gillespie, R. J.; Passmore, J. *Acc. Chem. Res.* **1971**, *4*, 413.
38. Borning, A. H.; Pullen, K. E. *Inorg. Chem.* **1969**, *8*, 1791.
39. Solomon, I. J.; Kacmarek, A. J.; Keith, J. N.; Raney, J. K. *J. Am. Chem. Soc.* **1968**, *90*, 6557.

# REACTIVITY OF HYDROGEN PEROXIDE, ALKYL HYDROPEROXIDES, AND PERACIDS

*To activate $O_2$ to be a more effective oxidant; reduce it to HOOH*

The reactivity of hydroperoxides is primarily dependent upon their unique bond energies (e.g., H-OOH, 90 kcal; HO-OH, 51 kcal; H-OOBu-*t*, 91 kcal, HO-OBu-*t*, 47 kcal),[1] which allow low-energy rearrangements to give (HO·) and (·O·).[2]

$$\text{HOOH} \longrightarrow \text{HOH} + [\text{O}], \Delta H, 39 \text{ kcal} \tag{4-1}$$

$$\text{HOOH} + \text{Fe}^{II}L_2 \longrightarrow L_2\text{Fe}^{III}\text{-OH} + \text{HO·}, \Delta H_{H_2O}, -2 \text{ kcal} \tag{4-2}$$

$$m\text{-ClPhC(O)OOH} \longrightarrow m\text{-ClPhC(O)OH} + [\text{O}], \Delta H, 31 \text{ kcal} \tag{4-3}$$

$$\text{HOOH} \xrightarrow{\text{Fe}^{III}\text{Cl}_3} \text{HOOH} + [\text{Cl}_3\text{Fe=O}], \Delta H_{MeCN}, -11 \text{ kcal} \tag{4-4}$$

Thus, $m$-ClPhC(O)OOH in acetonitrile will epoxidize olefins without a catalyst, but the presence of an iron complex accelerates the process by several orders of magnitude.[3]

In water the Brønsted acidity of HOOH and HO· are essentially the same ($pK_a$ = 11.8 and 11.9, respectively). Electrochemical reduction of HOOH in acetonitrile yields molecular hydrogen (rather than cleavage of the HO–OH bond)[4]

$$\text{HOOH} + e^- \xrightarrow[\text{GC, MeCN}]{-1.7 \text{ V versus SCE}} \tfrac{1}{2}\text{H}_2 + \text{HOO}^- \tag{4-5}$$

and oxidation of HOOH yields perhydroxyl radical (HOO·)

$$\text{HOH} \xrightarrow[\text{GC, MeCN}]{+2.1 \text{ V versus SCE}} \text{HOO·} + \text{H}^+ + e^- \tag{4-6}$$

Hydrogen peroxide is formed from (1) the dimerization of hydroxyl radicals,

$$2 \text{ HO·} \longrightarrow \text{HOOH}, k, 1 \times 10^9 \, M^{-1} \, s^{-1} \tag{4-7}$$

(2) the proton-induced decomposition of superoxide ions,

$$2 \text{ O}_2^{-\cdot} + 2 \text{ H}^+ \longrightarrow \text{HOOH} + \text{O}_2 \tag{4-8}$$

(3) the reductive electrolysis of $O_2$ in acidic media,

$$O_2 + 2\,H^+ + 2\,e^- \xrightarrow[\text{MeCN, Pt}]{+0.4\text{ V versus SCE}} HOOH \qquad (4\text{-}9)$$

and (4) the base-induced reduction of $O_2$ by 1,2-disubstituted hydrazines.[5]

$$O_2 + PhNHNHPh \xrightarrow{:B} HOOH + PhN=NPh \qquad (4\text{-}10)$$

Hydrogen peroxide is produced industrially by (1) the base/anthraquinone-catalyzed chemical reduction of $O_2$ by water and (2) the palladium-catalyzed hydrogenation of $O_2$. In aerobic biology many oxidases (dehydrogenases) produce HOOH as a biproduct along with oxidized substrate (e.g., *glucose oxidase*, $O_2$ + glucose $\xrightarrow{GO}$ gluconic acid + HOOH; *xanthine oxidase*, $O_2$ + xanthine $\xrightarrow{XO}$ uric acid + HOOH). Alkyl hydroperoxides and peracids are formed via the reaction of HOOH with alkyl halides and acid chlorides, respectively.

### Elementary reaction chemistry

Hydrogen peroxide is (1) a modest electron-transfer oxidant that requires a one-electron catalyst (usually iron or copper)

$$HOOH + 2\,I^- + 2\,HA \xrightarrow{\text{cat}} I_2 + 2\,HOH + 2\,A^- \qquad (4\text{-}11)$$

and (2) a weak two-equivalent reductant that also requires a one-electron catalyst.

$$2\,MnO_4^- + 5\,HOOH + 6\,H^+ \xrightarrow{\text{cat}} 5\,O_2 + 2\,Mn^{2+} + 8\,HOH \qquad (4\text{-}12)$$

The hydroperoxide ion ($HOO^-$) in aprotic media is an effective nucleophile that (1) oxygenates sulfoxides,[6]

$$HOO^- + Me_2SO \longrightarrow Me_2SO_2 + HO^- \qquad (4\text{-}13)$$

(2) hydrolyzes nitriles and amides,

$$HOO^- + MeC{\equiv}N \xrightarrow{2\,HOH} MeC{\overset{NH_2}{\underset{O^-}{-}}}OH + HOOH \qquad (4\text{-}14)$$

$$HOO^- + Me_2NCH(O) \xrightarrow{HOH} Me_2NH + HC(O)O^- + HOOH \qquad (4\text{-}15)$$

and (3) reacts with HOOH to form superoxide ions and hydroxyl radicals.[7]

$$\text{HOO}^- + \text{HOOH} \longrightarrow \text{O}_2^{-\cdot} + \text{HOH} + \cdot\text{OH} \qquad (4\text{-}16)$$

$$2\,\text{HOO}^- + \text{HOOH} \longrightarrow 2\,\text{O}_2^{-\cdot} + 2\,\text{HOH} \qquad (4\text{-}17)$$

**Fenton chemistry**

Although the activation of hydrogen peroxide by reduced transition metals has been known for almost 100 years as Fenton chemistry,[8]

$$\text{HOOH} + \text{LFe}^{II} \xrightarrow{pH\,2} \text{LFe}^{III}\text{-OH} + \text{HO}\cdot,\ k,\ {\sim}100\ M^{-1}\,\text{s}^{-1} \qquad (4\text{-}18)$$

$$\text{HO}\cdot + \text{LFe}^{II} \longrightarrow \text{LFe}^{III}\text{-OH},\ k,\ 2\text{x}10^8\ M^{-1}\,\text{s}^{-1} \qquad (4\text{-}19)$$

$$\text{HO}\cdot + RH \longrightarrow R\cdot + \text{H}_2\text{O},\ k,\ 10^7\text{-}10^{10}\ M^{-1}\,\text{s}^{-1} \qquad (4\text{-}20)$$

$$\text{HO}\cdot + \text{HOOH} \longrightarrow \text{HOO}\cdot + \text{H}_2\text{O},\ k,\ 1\text{x}10^7\ M^{-1}\,\text{s}^{-1} \qquad (4\text{-}21)$$

there is uncertainty as to whether this chemistry occurs within healthy aerobic organisms. Because free reduced iron is necessary to initiate this chemistry, its relevance to biology depends on the presence of soluble iron. The transition metals of metallo-proteins usually are buried within the protein matrix, which provides an environment that is distinctly different from that of bulk water (with respect to dielectric constant, proton availability, ionic solvation, and reduction potential). Hence, dipolar aprotic solvents such as dimethyl sulfoxide, acetonitrile, dimethylformamide, and pyridine/acetic acid more closely model the matrix of biological metals than does bulk water.

A recent study[9] has confirmed that the Fenton process [Eq. (4-18)] occurs in pyridine/acetic acid (2:1 molar ratio) when bis(picolinato)iron(II) [$\text{Fe}^{II}(\text{PA})_2$] and hydrogen peroxide are combined in 1:1 molar ratio (the rate constant for the $\text{Fe}^{II}(\text{PA})_2/\text{HOOH}$ reaction is $2\text{x}10^3\ M^{-1}\,\text{s}^{-1}$). In the presence of hydrocarbon substrates ($RH$) the HO· flux produces carbon radicals ($R\cdot$), which can be trapped by PhSeSePh to give $R$SePh. This chemistry, which is outlined in Scheme 4-1 for cyclohexane ($c$-$\text{C}_6\text{H}_{12}$), represents a reasonable model for HO· radical generation within biomembranes.

For the Fenton process to occur the $L_2\text{Fe}^{III}$-OH bond must have a $-\Delta G_{BF}$ value of at least 51 kcal [for $(\text{H}_2\text{O})_3^{2+}\,\text{Fe}^{III}$-OH (aqueous, $pH$ 2; $E°'$, + 0.4 V versus NHE) the value is 54 kcal; see Chapter 3]. At $pH$ 0 the potential for the $\text{Fe}^{III/II}(\text{H}_2\text{O})_6^{3+/2+}$ redox couple ($E°$, + 0.77 V versus NHE) indicates that the $(\text{H}_2\text{O})_5^{2+}\,\text{Fe}^{III}$-OH$_2^+$ bond has a $-\Delta G_{BF}$ value of 45 kcal, which is consistent with the fact that ferrous ion does not induce Fenton chemistry in the presence of 1 $M$ $\text{H}_3\text{O}^+$. In aqueous systems ($pH$ 2–10) and in the presence of a Brønsted base (pyridine in a py/HOAc solvent), the Fenton process is initiated via nucleophilic

*Scheme 4-1*   *Hydrocarbon selenization via Fenton chemistry in py/HOAc*

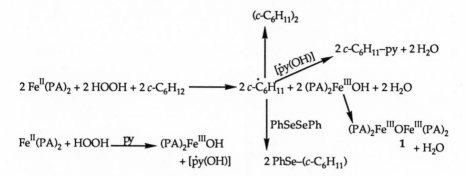

addition of HOOH to the iron(II) center, for example,

$$(PA)_2Fe^{II} + HOOH + py \rightarrow [(PA)_2^- \ Fe^{II}OOH + Hpy^+]$$

$$py + (PA)_2Fe^{III}OH + HO\cdot \qquad (4\text{-}22)$$

**Activation of HOOH by Lewis acids**

In a rigorously nonbasic solution matrix (e.g., dry acetonitrile) reduced iron without basic ligands (e.g., $[Fe^{II}(MeCN)_4](ClO_4)_2$) is not oxidized by hydroperoxides [HOOH, $t$-BuOOH, or $m$-ClPhC(O)OOH]. Table 4-1 summarizes the reactivity and products from the combination of iron(II) and hydrogen peroxide in MeCN with a series of organic substrates.[10,11] Whereas Fenton chemistry in aqueous solutions produces a diverse group of products via the radical processes that are induced by HO·, the products in a nonbasic matrix such as acetonitrile are characteristic of dehydrogenations, mono-oxygenations, and dioxygenations. That traditional Fenton chemistry does not occur is confirmed by the complete retention of the reduced oxidation state for iron.

The slow addition of dilute HOOH [pure HOOH (98%) in dry acetonitrile (MeCN)], $t$-BuOOH, or $m$-ClPhC(O)OOH to a solution that contains $[Fe^{II}(MeCN)_4](ClO_4)_2$ and an organic substrate ($RH$) in dry MeCN (<0.004% $H_2O$) results in the dehydrogenation (oxidation) or mono-oxygenation of $RH$. The net reactions and the conversion efficiencies (moles of substrate per mole of HOOH) for 1:0.5:0.5 $RH$/Fe(II)/HOOH combinations of several organic substrates are summarized in Table 4-2. The results from studies with $t$-BuOOH and $m$-ClPhC(O)OOH as the oxidant in place of HOOH also are included. In sharp contrast to aqueous Fe(II)/HOOH radical chemistry, the products from the $RH$/Fe(II)/ROOH combination in dry acetonitrile are characteristic of oxidase (dehydrogenation) and mono-oxygenation reactions. The total absence of

Table 4-1  Products from the Iron(II)-Induced Mono-oxygenation, Dehydrogenation, and Dioxygenation of Organic Substrates (RH) by HOOH in Dry Acetonitrile[a]

| Substrate | Reaction[b] efficiency (%) | Products |
|---|---|---|
| Mono-oxygenation | | |
| Blank (HOOH) | 100 | $O_2$, $H_2O$, Fe(II) |
| $Ph_3P$ | 100 | $Ph_3PO$ |
| $Me_2SO$ | 100 | $Me_2SO_2$ |
| $Ph_2SO$ | 100 | $Ph_2SO_2$ |
| EtOH | 70 | $MeCH(O)$ (90%), $MeC(O)OH$ (10%), $O_2$ |
| $PhCH_2OH$ | 100 | $PhCH(O)$ |
| Cyclohexanol | 47 | Cyclohexanone, $O_2$ |
| $MeCH(O)$ | 20 | $MeC(O)OH$, $O_2$ |
| $Me_2CO$ | NR | $O_2$ |
| $PhCH(O)$ | 28 | $PhC(O)OH$, $O_2$ |
| | | |
| Dehydrogenation and oxidation | | |
| Cyclohexane | NR | $O_2$ |
| 1,4-$c$-$C_6H_8$ | 59 | $PhH$, $O_2$ |
| PhNHNHPh | 100[c] | PhN=NPh |
| $H_2S$ | 100 | $H_2SO_4$ |
| $H_2O$ (56 mM) | 100 | Fe(III) |

## Dioxygenation

| Substrate | Yield[b] | Product |
|---|---|---|
| (structure: furan with Ph, Ph, Ph substituents) | 100 | (benzene ring) C(O)Ph, C(O)Ph |
| (anthracene-type structure with Ph, Ph, Ph, Ph) | 69 | (anthracene endoperoxide structure with Ph, Ph, O–O) |
| (pentacene-type structure with Ph, Ph, Ph, Ph) | 83 | (pentacene endoperoxide structure with Ph, Ph, Ph, Ph, O–O) |
| $Ph_2C=CPh_2$ | 22 | $Ph_2C(O)$, $O_2$ |
| $PhC≡CPh$ | 42 | $PhC(O)C(O)Ph$, $O_2$ |
| $PhC≡CMe$ | 26 | $PhC(O)C(O)Me$, $O_2$ |
| $PhC≡CH$ | 11 | $PhC(O)CH(O)$, $O_2$ |
| $c\text{-}PhCH=CHPh$ | 52 | $PhCH(O)$ (98%), $PhC≡CPh$ (2%), $O_2$ |
| $t\text{-}PhCH=CHPh$ | 28 | $PhCH(O)$, $O_2$ $t\text{-}PhCH=CHMe$ |
| $t\text{-}PhCH=CHM$ | 32 | $PhCH(O) + MeCH(O)$ (85%), $PhCHCHMe$ (15%), $O_2$ (epoxide structure) |

[a] Product solution [from the slow addition [~5 min to give a final 200 mM concentration] of 1 M HOOH [98% HOOH in MeCN] to a solution of 100 mM [Fe(MeCN)$_4$](ClO$_4$)$_2$ plus 200 mM substrate] analyzed by gas chromatography and assayed for residual Fe(II) by colorimetry with MnO$_4^-$ titration and by colorimetry with 1,10-$o$-phenanthroline.

[b] 100% represents one substrate oxygenation or dehydrogenation per HOOH added. For dioxygenations, 100% represents one substrate converted per two HOOH added.

Table 4-2  Conversion Efficiencies for Fe(II)-ROOH [R = H, t-Bu, m-ClPhC(O)] Induced Dehydrogenations and Mono-oxygenations of Organic Substrates (RH) in Acetonitrile[a]

| Substrate reaction | HOOH[c] | t-BuOOH | m-ClPhC(O)OOH |
|---|---|---|---|
| a. Dehydrogenations | | | |
| HOOH → $O_2$ | 100 | | |
| 1,4-$c$-$C_6H_8$ → PhH | 59 | 100 | 53 |
| 1,3-$c$-$C_6H_8$ → [PhH/$(C_6H_7)_2$] | 86 [1/2] | 89 [1/2] | 41 [1/2.5] |
| PhNHNHPh → PhN=NPh | 100 | 75 | 100 |
| 3,5-$(t$-Bu$)_2$-1,2-$(OH)_2C_6H_2$ → 3,5-$(t$-Bu$)_2$-$o$-benzoquinone | 100 | 70 | 90 |
| 2PhCH$_2$SH → PhCH$_2$SSCH$_2$Ph | 34/2 | 32/2 | 10/2 |
| 2PhSH → PhSSPh | 68/2 | 32/2 | 10/2 |
| b. Mono-oxygenations | | | |
| $c$-$C_6H_{11}$OH → $c$-$C_6H_{10}$(O) | 47 | 27 | 45 |
| MeCH$_2$OH → MeCH(O) | 70 | 20 | 10 |
| PhCH$_2$OH → PhCH(O) | 100 | 72 | 48 |
| PhCH$_2$OBu-$t$ → PhCH(O) + $t$-BuOH | 30 | 70 | 60 |
| MeCH(O) → MeC(O)OH | 20 | 10 | 27 |

| | 28 | 9 | 28 |
|---|---|---|---|
| PhCH(O) → PhC(O)OH | 28 | | 28 |
| t-PhCH=CHMe → PhCH—CHPh /[PhCH(O) + MeCH(O)] (epoxide) | 16 [1/3] | 13 [1/4] 29 [9/1] | 29[9/1] |
| PhCH—CHMe (epoxide) → dioxane (dimer) + [PhCH(O) + MeCH(O)] | 80 [7/3] | 80[4/1] | 85 [19/1] |
| $PhCH_3$ → $PhCH_2OH$ [+PhCH(O), $CH_3Ph(OH)$] | 0.3 | 2 | 0 |
| $(c\text{-}C_6H_{11})_2S$ → $(c\text{-}C_6H_{11})_2S(O)$ | 35 | 0 | 15 |
| $Ph_2S$ → $Ph_2SO$ | 27 | 7 | 51 |
| $Ph_2SO$ → $Ph_2S(O)_2$ | 100 | 1 | 71 |
| $Ph_3P$ → $Ph_3PO$ | 100 | 100 | 47 |

a To 1.0 mmol of substrate and 0.5 mmol of $[Fe^{II}(MeCN_4](ClO_4)_2$ in 10 mL of MeCN was added slowly 0.5 mmol of hydroperoxide (1 M ROOH in MeCN). Reaction time and temperature: 23°C for 5 min [$1,4\text{-}c\text{-}C_6H_8$, $1,3\text{-}c\text{-}C_6H_8$, PhNHNHPh, $PhCH_2SH$, PhSH, $c\text{-}C_6H_{11}OH$, $CH_3CH(O)$, PhCH(O), $PhCH_3$, $Ph_2SO$ and $Ph_3P$]; 5°C for 10 min [$3,5\text{-}(t\text{-}Bu)_2\text{-}1,2\text{-}(OH)_2C_6H_2$, $C_2H_5OH$, $PhCH_2OH$, $PhCH_2OCMe_3$, PhCH=CHMe, PhCH—CHMe , $(c\text{-}C_6H_{11})_2S$, and $Ph_2S$].

b 100% represents one substrate dehydrogenation or oxygenation per ROOH added to the reaction system. In the case of HOOH all of it was consumed, either by oxygenation of substrate or by disproportionation to $O_2$ and $H_2O$. Starting material was recovered for those substrates that react with less than 100% efficiency.

c These numbers represent a crude measure of the relative extent of reaction for the adduct with substrate and HOOH.

products from (HO·) radical chemistry, and of any Fe(III) in the product solutions, confirms that classical Fenton chemistry does not occur.[1]

In the absence of substrate, the combination of Fe(II) and HOOH in dry MeCN results in the latter's rapid disproportionation to $O_2$ and $H_2O$, but the Fe(II) remains unoxidized. Similar combinations of Fe(II) with $t$-BuOOH or m-ClPhC(O)OOH do not promote their rapid disproportionation, but facilitate a slow hydrolysis process ($ROOH + H_2O \rightarrow ROH + HOOH$, assayed for $ROH$).

Because all of the HOOH is consumed in the experiments of Table 4-2, the reaction efficiency for the Fe(II)-HOOH system represents a crude measure of the relative rate of reaction for the adduct with substrate and HOOH. The relative order of the reaction rates for the substrates of Table 4-2 has been determined via a series of competition experiments with equimolar amounts of $RH$ and $Ph_2SO$. Hence, the extent of oxidative conversion of $RH$ relative to $Ph_2SO$ from the slow addition of dilute HOOH (0.5 mmol) to an MeCN solution that contains 0.5 mmol of $RH$, 0.5 mmol of $Ph_2SO$, and 0.5 mmol of Fe(II) has been measured; the results are summarized in Table 4-3 as relative amounts reacted. The bond energies for the weakest $R$–$H$ bond of the substrate also are tabulated.[1]

In general, the [Fe(II)/$ROOH$]-induced dehydrogenations and mono-oxygenations for the substrates in Table 4-2 exhibit a first-order dependence on the $RH$ and $ROOH$ concentrations for their reaction efficiencies (and reaction rates). The presence of $O_2$ (at 1 atm) enhances the conversion efficiency for PhCH(O) by a factor of 370 and causes all of the Fe(II) catalyst to be oxidized, but has a negligible effect on the oxygenation of $PhCH_2OH$. This result indicates that the Fe(II)/HOOH reagent can induce the auto-oxygenation of PhCH(O).

Scheme 4-2 outlines an activation cycle that is based on the argument that within an aprotic matrix the iron(II) center is a strong Lewis acid that weakens the O–O bond of electron transfer from the iron(II).[12] The products for the Fe[II]/HOOH oxidations are consistent with those that result from some peroxidase-catalyzed processes.

*Iron(II)-induced activation of hydroperoxides for the dehydrogenation of organic substrates*

The results of Table 4-2 confirm that the Fe[II]($ROOH$)$^{2+}$ adducts are effective dehydrogenation agents for substrates such as cyclohexadienes, substituted hydrazines, catechols, and thiols. For PhNHNHPh and 3,5-($t$-Bu)$_2$-1,2-$(OH)_2C_6H_2$ the reaction efficiencies are comparable for HOOH and $m$-ClPhC(O)OOH as the oxidant but somewhat reduced for $t$-BuOOH. This may indicate that the oxene complex (**2**, Scheme 4-2) is the dominant reactive form and that $t$-BuOOH is hindered from assuming this configuration. Thus, the end-on configuration (**2**) with its oxene character is favored for HOOH and $m$-ClPhC(O)OOH.

1,4-cyclohexadiene yields only benzene for the three Fe[II]($ROOH$)$^{2+}$ adducts. Either the bound substrate induces the homolytic scission of the $RO$–OH bond to give bound $RO$· and ·OH (**1**, Scheme 4-2) for the concerted removal of two H atoms from the substrate or the oxenelike character of the end-on configuration (**2**, Scheme 4-2) results in the same concerted removal of two H

Table 4-3  Relative Extent of Reaction for $[Fe(II)/HOOH]^{2+}$-Induced Dehydrogenations and Mono-oxygenations of Organic Substrates ($RH$) in Acetonitrile[a]

| Substrate reaction | Relative amt reacted[b] ($RH/Ph_2SO$) | $R–H$ bond energy[c] (kcal) |
|---|---|---|
| a.  Dehydrogenations | | |
| $1,4$-$c$-$C_6H_8 \rightarrow PhH$ | 8.7 | 73 |
| $1,3$-$c$-$C_6H_8 \rightarrow PhH + (c$-$C_6H_7)_2$ | 8.7 | 73 |
| $PhNHNHPh \rightarrow PhN=NPh$ | 5.4 | |
| $2PhCH_2SH \rightarrow PhCH_2SSCH_2Ph$ | 1.3 (2.6/2) | 86 |
| $2\,PhSH \rightarrow PhSSPh$ | 2.2 (4.4/2) | 83 |
| b.  Mono-oxygenations | | |
| $c$-$C_6H_{11}OH \rightarrow c$-$C_6H_{10}(O)$ | 0.12 | 93 |
| $MeCH_2OH \rightarrow MeCH(O)$ | 0.17 | 93 |
| $PhCH_2OH \rightarrow PhCH(O)$ | 0.27 | 83 |
| $PhCH_2OBu$-$t \rightarrow PhCH(O)$ | 0.32 | 83 |
| $MeCH(O) \rightarrow MeC(O)OH$ | 0.48 | 86 |
| $PhCH(O) \rightarrow PhC(O)OH$ | 0.57 | 87 |
| $PhCH=CHMe \rightarrow PhCH\overset{O}{-}CHMe$ +  $PhCH(O) + MeCH(O)$ | 0.54 | |
| $(c$-$C_6H_{11})_2S \rightarrow (C_6H_{11})_2SO$ | 0.55 | |
| $Ph_2S \rightarrow Ph_2SO$ | 0.26 | |
| $Ph_2SO \rightarrow Ph_2SO_2$ | 1.00 | |
| $Ph_3P \rightarrow Ph_3PO$ | 16.1 | |

[a] To the acetonitrile solution (10 mL) of the substrates (0.5 mmol each) and $[Fe^{II}(MeCN)_4](ClO_4)_2$ (0.5 mmol) was slowly added 1 $M$ HOOH (0.5 mmol) under argon at 23°C. The reaction mixture was stirred for 5 min.  After extraction of the product solution with ether, the relative extent of conversion was determined on the basis of the residual substrates as determined by gas chromatography.

[b] The numbers represent the relative extent for the oxidative conversion from the addition of 0.5 mmol of HOOH to 0.5 mmol of $RH$ and 0.5 mmol $Ph_2SO$ in 10 mL of MeCN.

[c] Ref. 1.

*Scheme 4-2   Activation of ROOH by iron(II) Lewis acids*

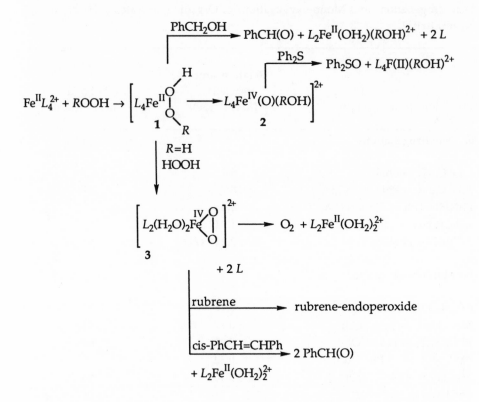

atoms.   For the 1:1 1,3-cyclohexadiene/$Fe^{II}(ROOH)^{2+}$ systems, a major fraction of the product is the $(c\text{-}C_6H_7)_2$ dimer, especially for high substrate–Fe(II) ratios with HOOH or $m$-ClPhC(O)OOH as the oxidants (Table 4-2).   Although $H_2S$ is oxygenated to $H_2SO_4$ by the $Fe^{II}(HOOH)^{2+}$ adduct (four HOOH molecules per $H_2S$),[11] thiols (both aromatic and aliphatic) are dehydrogenated by the $Fe^{II}(ROOH)^{2+}$ complex to give disulfides as the only product.   The reaction stoichiometry of two $RSH$ molecules per $ROOH$ is the same as for the dimerization of 1,3-$c$-$C_6H_8$, a similar bis adduct for the reaction complex probably is formed [Eq. (4-23)].

$$2\,RSH + Fe(II) + ROOH \rightarrow [(RSH)_21]^{2+} \rightarrow RSSR + Fe^{II}(H_2O)(ROH)^2 \quad (4\text{-}23)$$

Because the sideon configuration (**1**, Scheme 4-2) of the $Fe(II)(ROOH)^{2+}$ adduct appears to be the reactive form, the lower yields for the $m$-ClPhC(O)OOH oxidant indicate that the oxene configuration (**2**, Scheme 4-2) is dominant for the peracid. The chemistry for the reaction of Eq. (4-22) is equivalent to that of glutathione peroxidase (an enzyme that contains selenocysteine in its active site),[13] and the $Fe^{II}(ROOH)^{2+}$ complex may represent a reaction mimic for the enzyme.

There is a rough inverse correlation between the relative reaction rates that are summarized in Table 4-3 and the weakest $R$–H bond energy of the substrates. This is in accord with the biradical and homolytic bond-scission mechanisms that are proposed in Scheme 4-2.

*Iron(II)-induced activation of hydroperoxides for the mono-oxygenation of organic substrates*

The results of Table 4-2 establish that the $Fe^{II}(ROOH)^{2+}$ adducts in dry MeCN are effective mono-oxygenases for alcohols (and for activated olefins, thioethers, sulfoxides, and phosphines). The reaction rate for $PhCH_2OH$ is about twice as fast as for $MeCH_2OH$ (Table 4-3), probably because the C–H bond energy for the methylene group of the former is 10 kcal smaller. Also, the reaction rates for a given alcohol are in the order $HOOH > t\text{-}BuOOH > m\text{-}ClPhC(O)OOH$ (Table 4-2), and the rate of oxygenation for $PhCH_2OBu\text{-}t$ by the $Fe^{II}(t\text{-}BuOOH)^{2+}$ adduct is essentially the same as that for $PhCH_2OH$ (ratio of reactivity: $PhCH_2OH:PhCH_2OBu\text{-}t = 1.0:1.2$).

These results indicate that the $Fe^{II}(ROOH)^{2+}$ adduct (Scheme 4-2) oxygenates alcohols (and ethers) and that HOOH and $t\text{-}BuOOH$ are more reactive than $m\text{-}ClPhC(O)OOH$. A mechanism that is consistent with these observations involves either (1) the homolytic scission of the sideon $RO–OH$ bond [1, Scheme 4-2; induced by the bound substrate $(ROH)$] and the subsequent abstraction by $RO\cdot$ of an H atom from the α-carbon or (2) the direct abstraction by the oxene oxygen of the endon configuration (2, Scheme 4-2) of an H atom from the α-carbon and the subsequent addition of the resulting $\cdot OH$ group to the carbon radical [Eq. (4-24)].

$$PhCH_2OH + Fe(II) + ROOH$$

$$\rightarrow \left[ \underset{\cdot}{(PhC\overset{H}{H}O:)}Fe(\cdot OH)(ROH) \rightarrow [PhCH(OH)_2]Fe^{II}(ROH) \right]^{2+} \quad (4\text{-}24)$$

$$\longrightarrow PhCH(O) + Fe^{II}(OH_2)(ROH)^{2+}$$

The resulting hemiacetal dissociates to give the aldehyde and $H_2O$ [$R'OH$ when the substrate is an ether $(ROR')$]. Because ethers are as reactive as alcohols and give the same aldehyde product, the process must be a mono-oxygenation to the hemiacetal rather than a dehydrogenation (concerted removal of hydrogen atoms from the α-carbon and the OH group).

Reference to the data of Tables 4-2 and 4-3 confirms that the $Fe^{II}(ROOH)^{2+}$ activated complexes from HOOH, $t\text{-}BuOOH$, and $m\text{-}ClPhC(O)OOH$ are effective mono-oxygenases for aldehydes. The much higher rates when HOOH and $m\text{-}ClPhC(O)OOH$ are the oxidants relative to those for $t\text{-}BuOOH$ may indicate that the oxene character of the endon configuration (2, Scheme 4-2) makes it the dominant reactive complex [Eq. (4-25)].

$$PhCH(O) + Fe(II) + ROOH \rightarrow \{[PhCH(O)]2 \rightarrow [Ph\overset{\cdot}{C}(O)] [Fe(\cdot OH)(ROH)]\}^{2+}$$

$$\rightarrow PhC(O)OH + Fe^{II}(ROH)^{2+} \quad (4\text{-}25)$$

The HOOH and *m*-ClPhC(O)OOH adducts of Fe(II) exhibit much higher reactivity relative to that for *t*-BuOOH for substrates that undergo an O-atom addition rather than a biradical process. Apparently, the $Fe^{II}(t\text{-BuOOH})^{2+}$ adduct has a limited tendency to take on significant oxene character in the endon configuration (**2**, Scheme 4-2).

The dramatic enhancement by molecular oxygen of the rate and extent of the $Fe^{II}(HOOH)^{2+}/PhCH(O)$ reaction is indicative of an auto-oxygenation process. Thus, the formation of the reactive intermediate complex provides a biradical center that can couple with triplet oxygen ($\cdot O_2 \cdot$) to give $PhC(O)OO\cdot$ in an initiation step [(Eq. 4-26)].

*Initiation*:

$$PhCH(O) + Fe(II) + HOOH \rightarrow \{[PhCH(O)]Fe^{II}(HOOH) \rightarrow$$

$$[Ph\overset{\cdot}{C}(O)] [Fe(OH)(OH_2)]\}^{2+} \xrightarrow{\cdot O_2\cdot} PhC(O)OO\cdot + Fe^{III}(OH)(OH_2)^{2+} \qquad (4\text{-}26)$$

The coupling by $\cdot O_2 \cdot$ with the carbon radical leaves an $[Fe(\cdot OH)]$ center, which goes to the observed $Fe^{III}(OH)(OH_2)^{2+}$ product. The peroxy radical $[PhC(O)OO\cdot]$ from the initiation step apparently abstracts an H atom from a second PhCH(O), and the resulting $Ph\overset{\cdot}{C}(O)$ radical couples with another $\cdot O_2 \cdot$ in the propagation step [Eq. (4-27)].

*Propagation*:

$$PhC(O)OO\cdot + PhCH(O) \rightarrow PhC(O)OOH + Ph\overset{\cdot}{C}(O)$$

$$\overset{\displaystyle \lfloor \quad O_2}{\phantom{x}} \xrightarrow{\phantom{xxx}} PhC(O)OO\cdot \qquad (4\text{-}27)$$

The Fe(II)-catalyzed oxygenation of another PhCH(O) by the peracid represents a second propagation step [Eq. (4-28)].

$$PhCH(O) + PhC(O)OOH \xrightarrow{Fe(II)} 2\,PhC(O)OH \qquad (4\text{-}28)$$

The sum of these processes is an $Fe^{II}(HOOH)^{2+}$-catalyzed auto-oxygenation of PhCH(O) [Eq. (4-29)].

$$2\,PhCH(O) + O_2 \xrightarrow{Fe^{II}(HOOH)^{2+}} 2\,PhC(O)OH \qquad (4\text{-}29)$$

The mono-oxygenation of methylstyrene (PhCH=CHMe) to form the epoxide (Table 4-2) appears to involve an O-atom transfer from the endon configuration (**2**, Scheme 4-2) of the $Fe^{II}(ROOH)^{2+}$ complex.

$$PhCH=CHMe + Fe(II) + ROOH \rightarrow [(PhCH=CHMe)2]^{2+} \rightarrow \overset{\diagup O \diagdown}{PhCH-CHMe}$$
$$+ Fe^{II}(ROH)^{2+} \quad (4\text{-}30)$$

However, a significant fraction of the products from this substrate (75% for HOOH and 80% for $t$-BuOOH, Table 4-2) is the result of a dioxygenation to give PhCH(O) and MeCH(O). When $m$-ClPhC(O)OOH is the oxidant, only 10% of the PhCH=CHMe that reacts is dioxygenated. These results are consistent with the proposition that the endon configuration of the $Fe^{II}[m\text{-}ClPhC(O)OOH]^{2+}$ adduct has the most oxene character and favors O-atom transfer to PhCH=CHMe and that the $Fe^{II}(HOOH)^{2+}$ and $Fe^{II}(t\text{-}BuOOH)^{2+}$ adducts react via a biradical mechanism. A reasonable mechanistic pathway that involves two $ROOH$ molecules per PhCH=CHMe molecule is presented in Eq. (4-31).

$$PhCH=CHMe + L_4^{2+}Fe(II) + ROOH \underset{2 L}{\longrightarrow} \left[ L_2(ROH)_2^{2+}Fe^{IV}\overset{\diagup O}{\underset{\diagdown O}{|}} + PhCH=CHMe \right]$$

$$\downarrow$$

$$PhCH(O) + MeCH(O) + L_2Fe^{II}(ROH)_2^{2+} \quad (4\text{-}31)$$

The latter process is dominant when the $ROOH$ concentration is greater than that for the Fe(II) catalyst. When the oxidant is HOOH and is in excess relative to Fe(II) and substrate, the disproportionation of HOOH is favored via an activated dioxygen intermediate (**3**, Scheme 4-2), which dioxygenates aromatic olefins [e.g., PhCH=CHPh $\rightarrow$ 2 PhCH(O)].[11] Hence, PhCH=CHMe probably is subject to dioxygenation by this activated intermediate [Eq. (4-32)].

$$PhCH=CHMe + Fe(II) + 2 HOOH \rightarrow$$

$$\left[ (PhCH=CHMe) \, \mathbf{3} \rightarrow \overset{O-O}{\underset{|\quad|}{(PhCH-CHMe)Fe^{II}(OH_2)_2}} \right]^{2+}$$

$$\rightarrow PhCH(O) + MeCH(O) + FeII(OH_2)_2^{2+} \quad (4\text{-}32)$$

The limited reactivity of the $Fe^{II}(t\text{-}BuOOH)^{2+}$ adduct with thioethers ($R_2S$) and sulfoxides ($R_2SO$) (Table 4-2) probably is due to its limited oxene character. The simplest pathway to the mono-oxygenated products for these substrates is via O-atom transfer from the endon complex (**2**, Scheme 4-2)

$$Ph_2S + Fe(II) + ROOH \rightarrow [(Ph_2S)2]^{2+} \rightarrow Ph_2SO + Fe^{II}(ROH)^{2+} \quad (4\text{-}33)$$

$$Ph_2SO + Fe(II) + ROOH \rightarrow [(Ph_2SO)2]^{2+} \rightarrow Ph_2SO_2 + Fe^{II}(ROH)^{2+} \quad (4\text{-}34)$$

with the bound substrates inducing the heterolytic formation of a reactive oxygen atom from the bound $ROOH$ group [$R$ = H or $m$-ClPhC(O)].  Phosphines (e.g., Ph₃P) probably are mono-oxygenated by a mechanism analogous to that for Ph₂S [Eq. (4-33)].

The formulations of Scheme 4-2 for the reactive forms of the $Fe^{II}(ROOH)^{2+}$ adducts represent unique electrophilic centers that are consistent with reasonable dehydrogenase (1), mono-oxygenase (1 and 2), and dioxygenase (3) reaction mechanisms.  The mono-oxygenase formulations also are consistent with the redox stoichiometry of the cytochrome $P$-450 cycle[13] and represent a form of oxygen activation that promotes electrophilic abstraction of a hydrogen atom (or O-atom transfer) to give radical (or oxene) activated intermediates[14] and the mono-oxygenation of cytochrome $P$-450 substrates (Table 4-2).[15]

In the presence of excess HOOH the $Fe^{II}(MeCN)_4{}^{2+}$ catalyst forms a

reactive adduct, $\left[ L_2(H_2O)_2Fe \overset{IV}{\underset{O}{\overset{O}{<}}} \right]^{2+}$ , that reacts with diphenylbenzofuran, aryl

olefins, 9,10-diphenylanthracene, or rubrene to form exclusively dioxygenated products (Table 4-1 and Scheme 4-2).[11] Such reactivities parallel those of dioxygenases with this group of substrates.

The $[Fe^{II}(OPPh_3)_4](ClO_4)_2$ complex also activates HOOH in a manner similar to $[Fe^{II}(MeCN)_4](ClO_4)_2$, but, in contrast, forms a stable binuclear product [(Eq. 4-35)] that is able to dioxygenate PhC≡CPh.[16]

$$2\,Fe^{II}(OPPh_3)_4{}^{2+} + 2\,HOOH \rightarrow [(Ph_3PO)_4(H_2O)Fe^{III}OOFe^{III}(OH_2)(OPPh_3)_4]^{4+}$$

$$(4\text{-}35)$$

### $Fe^{III}Cl_3$-induced epoxidation, mono-oxygenation, and dehydrogenation

In a base-free medium (dry MeCN) $Fe^{III}Cl_3$ activates HOOH to form a reactive intermediate that oxygenates alkanes, alkenes, and thioethers, and dehydrogenates alcohols and aldehydes.[17] Table 4-4 summarizes the conversion efficiencies and product distributions for a series of alkene substrates subjected to the $Fe^{III}Cl_3$/HOOH/MeCN system.  The extent of the $Fe^{III}Cl_3$-induced mono-oxygenations is enhanced by higher reaction temperatures and increased concentrations of the reactants (substrate, $Fe^{III}Cl_3$, and HOOH).  For 1-hexene (representative of all of the alkenes) a substantial fraction of the product is the dimer of 1-hexene oxide, a disubstituted dioxane

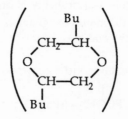

Table 4-4 Products and Conversion Efficiencies for the $Fe^{III}Cl_3$-Induced Oxygenation–Dehydrogenation of Olefins and Organic Substrates (RH) by HOOH in Dry Acetonitrile[a]

| Substrate | Reaction efficiency (percentage)[b] | Products[c] |
|---|---|---|
| a. Olefins (-5°C, 10-min reaction times) | | |
| blank (HOOH) | 100 | $O_2$, $H_2O$ |
| 1-hexene | 10 | epoxide (1-hexene oxide) (71%), dimer (dioxane) (10%), others (19%) |
| 1-hexene (+5°C) | 23 | epoxide (55%), dimer (15%), others (30%) |
| 1-octene | 60 | epoxide (53%), dimer (10%) |
| cyclohexene | 25 | epoxide (45%), dimer (30%) |
| norbornene | 52 | exo-epoxide (80%), nonepoxide products (20%) |
| 1,4-cyclohexadiene | 39 | PhH (76%), epoxide (17%) |
| cis-stilbene | 63 | PhCH(O) (50%), epoxides (50%) (cis:trans ratio, 2.5:1) |
| b. Other substrates (+5°C, 20-min reaction times) | | |
| cyclohexanol | 52 | cyclohexanone (88%) |
| $PhCH_2OH$ | 63 | PhCH(O) (51%), $PhCH_2Cl$ (21%), PhC(O)OH (14%), PhC(O)Cl (14%) |
| $PhCH_2OBu$-$t$ | 56 | PhCH(O) (72%), $PhCH_2Cl$ (11%), PhC(O)OH (3%), PhC(O)Cl (14%) |
| PhCH(O) | 75 | PhC(O)OH (55%), PhC(O)Cl (45%) |
| $PhCH_3$ (25°C) | 2 | $PhCH_2OH$, PhCH(O), PhC(O)Cl, PhC(O)OH, cresols |
| cyclohexane | 22 | cyclohexylchloride (45%), cyclohexanol (40%), cyclohexanone (15%) |
| PhC(Me)(OH)C(Me)(OH)Ph | 30 | PhC(O)Me (100%) |
| $PhNMe_2$ | 39 | PhNHMe (95%), PhN[CH(O)]Me (5%) |
| $Ph_2S$ | 58 | $Ph_2SO$ (100%) |
| $Ph_2SO$ | 60 | $Ph_2S(O)_2$ (100%) |
| $Ph_3P$ | 80 | $Ph_3PO$ (100%) |

[a] RH and $Fe^{III}Cl_3$ (1.0 mmol of each) combined in 10–20 mL of dry MeCN, followed by the slow addition of 1 mmol HOOH [1 M HOOH (98%) in MeCN].

[b] Percentage of substrate converted to products.

[c] After the indicated reaction time, the product solution was quenched with water, extracted with diethylether and analyzed by capillary gas chromatography and GC/MS.

With other organic substrates (RH), $Fe^{III}Cl_3$ activates HOOH for their mono-oxygenation; the reaction efficiencies and product distributions are summarized in Table 4-4b. In the case of alcohols, ethers, and cyclohexane, a substantial fraction of the product is the alkyl chloride, and with aldehydes [e.g., PhCH(O)] the acid chloride represents one-half of the product. In the absence of substrate the $Fe^{III}Cl_3$/MeCN system catalyzes the rapid disproportionation of HOOH to $O_2$ and $H_2O$.

Because $Fe^{III}Cl_3$ is an exceptionally strong Lewis acid and electrophilic center, it activates HOOH (which acts as a nucleophile) for the dehydrogenation of a second HOOH. On the basis of the disproportionation process, as well as the mono-oxygenation and dehydrogenation reactions of Table 4-4, the activation of HOOH by $Fe^{III}Cl_3$ probably involves the initial formation of at least two reactive forms of an $Fe^{III}Cl_3$ (HOOH) adduct (4 and 5, Scheme 4-3).

*Scheme 4-3 Proposed pathways for activation of HOOH by $Fe^{III}Cl_3$ in MeCN*

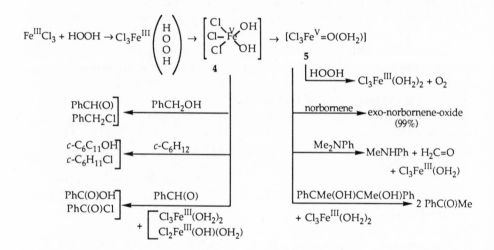

The disproportionation of HOOH occurs via a concerted transfer of the two hydrogen atoms from a second HOOH to the $Fe^{III}Cl_3$(HOOH) adduct. This dehydrogenation of HOOH is a competitive process with the $Fe^{III}Cl_3$/substrate/HOOH reactions. The controlled introduction of dilute HOOH into the $Fe^{III}Cl_3$/substrate solution limits the concentration of HOOH and ensures that the substrate/HOOH reaction can be competitive with the second-order disproportionation process. The substrate reaction efficiencies in Table 4-4 are proportional to the relative rates of reaction ($k_{RH}/k_{HOOH}$). The mode of activation of HOOH by $Fe^{III}Cl_3$ is analogous to that of $Fe^{II}(MeCN)_4^{2+}$; both are strong electrophiles in ligand-free dry MeCN and induce HOOH to mono-oxygenate organic substrates.

The epoxidation of alkenes (Table 4-4a) appears to involve an O-atom transfer from the oxene configuration of the $Fe^{III}Cl_3$(HOOH) adduct (5, Scheme 4-3). The resulting epoxides are rapidly dimerized to dioxanes. Hence, the complete conversion of an alkene to its epoxide is precluded; the more complete the conversion, the higher the fraction of dioxane in the product mixture.

The results in Table 4-4b indicate that the $Fe^{III}Cl_3(HOOH)$ adduct monooxygenates alkanes, alcohols, and aldehydes. A mechanism that is consistent with this involves the homolytic scission of the HO–OH bond in the sideon configuration (4, Scheme 4-3), induced by the bound substrate, and the subsequent abstraction by one ·OH of an H atom from the α-carbon and addition of the second ·OH to the resulting carbon radical (Scheme 4-3). An analogous process appears to occur for the oxygenation of benzaldehyde by the $Fe^{III}Cl_3(HOOH)$ adduct, but 50% of the product is the acid chloride. Occurrence of the latter indicates that the activated sideon complex (4, Scheme 4-3) has some hypochlorous acid (HOCl) character and can add a chlorine atom to the carbon radical that results from the H-atom abstraction by the ·OH group. This also occurs with alkanes, alcohols, and ethers (Table 4-4b). Such chemistry is similar to the activation of chloride ion and HOOH to HOCl by a heme protein, myeloperoxidase.[18,19] Phosphines, dialkylsulphides, and sulfoxides are monooxygenated by the $Fe^{III}Cl_3(HOOH)$ adduct in a manner that appears to be analogous to that for the epoxidation of alkenes via species 5 of Scheme 4-3.

Anhydrous $Fe^{III}Cl_3$ catalyzes the stereospecific epoxidation of norbornene, the demethylation of N,N-dimethylaniline, and the oxidative cleavage of PhCMe(OH)CMe(OH)Ph (and other α-diols) by hydrogen peroxide (Table 4-4 and Scheme 4-3).[17] For each class of substrate the products parallel those that result from their enzymatic oxidation by cytochrome *P*-450. The close congruence of the products indicates that the reactive oxygen in the $Fe^{III}Cl_3/HOOH$ model system and in the active form of cytochrome *P*-450 is essentially the same, with strong electrophilic oxene character (stabilized singlet atomic oxygen).

The reaction of the $Fe^{III}Cl_3/HOOH$ system with substrate is first order in both $Fe^{III}Cl_3$ and HOOH concentrations but independent of substrate concentration.[20] Such behavior is consistent with the rate-determining formation of a reactive intermediate from $Fe^{III}Cl_3/HOOH$. In the absence of direct structural evidence, conclusions as to the form of the intermediate are speculative. However, without electron transfer between the Fe(III) center and HOOH, the initial interaction between $Fe^{III}Cl_3$ and HOOH is that of a Lewis acid with a base, a process facilitated by the base-free solvent system. The deactivation of the $Fe^{III}Cl_3/HOOH$ system by donor ligands such as water or excess chloride ($Fe^{III}Cl_4^-$ is inert with respect to HOOH disproportionation or activation) also is consistent with a neutralization of the Lewis acidity of the Fe(III) center. The ability of $Fe^{III}Cl_3$ to activate peracids in a manner that parallels its activation of HOOH [coupled with the similar reactivity of iodosylbenzene induced by simple iron(III) salts in acetonitrile][21] indicates that a common reactive iron–oxygen intermediate is formed (5, Scheme 4-3).

The high degree of electrophilicity of the oxene intermediate would facilitate hydrogen-atom abstraction from substrates such as the methyl group of N,N-dimethylaniline to generate a "crypto-hydroxyl" metal center able to undergo the well-known "oxygen rebound" mechanism.[22]

$$Cl_3Fe^V{=}O + (CH_3)_2NPh \longrightarrow \left[ Cl_3Fe^{IV}{-}OH + \begin{array}{c} {\cdot}CH_2 \\ \diagdown \\ CH_3 \diagup \end{array} NPh \right] \longrightarrow$$

$$Cl_3Fe^{III} + \begin{array}{c} HOCH_2 \\ \diagdown \\ CH_3 \diagup \end{array} NPh \quad \overset{\qquad}{\underset{\qquad}{\Bigg|}} \longrightarrow CH_3NHPh + CH_2(O) \qquad (4\text{-}36)$$

Although a similar rebound process may occur in the cleavage of 1,2-diols [via the sequential abstraction of hydrogen atoms from carbon centers by the $Cl_3Fe^V(O)$ reactive intermediate], a more plausible process is the concerted removal of the two diol hydrogen atoms via intermediate **4** of Scheme 4-3 to generate a dioxetane intermediate, followed by a facile homolytic cleavage to the observed products [Eq. (4-37)].

$$Fe^{III}Cl_3 + HOOH + RCH(OH)CH(OH)R' \xrightarrow{\ MeCN\ } Cl_3Fe^{III}(OH_2)_2 + \left[ \begin{array}{c} O{-}O \\ | \quad | \\ RCHCHR' \end{array} \right]$$

$$\Big\downarrow$$

$$RCH(O) + R'CH(O) \qquad (4\text{-}37)$$

The $Fe^{III}Cl_3/HOOH/MeCN$ system is able to catalyze a variety of substrate transformations analogous to those of cytochrome *P*-450. The simplicity of the system allows the mechanism of the various reactions to be examined in the absence of the complications that are introduced by the electronic effects of the porphyrin ring and the axial thiolate group.

### Oxygen-atom transfer chemistry; formation and reactivity of atomic oxygen from hydroperoxides [HOOH, ROOH, R'C(O)OOH]

The function of peroxidase enzymes is the activation of HOOH to provide two oxidizing equivalents for the subsequent oxidation of a variety of substrates. The interaction of horseradish peroxidase [HRP, an iron(III)-heme that has a proximal imidazole] with HOOH results in the formation of a green reactive intermediate known as Compound I. The latter is reduced by one electron to give a red reactive intermediate, Compound II.[23] Both of these intermediates contain a single oxygen atom from HOOH, and Compound I is two oxidizing equivalents above the iron(III)-heme state with a magnetic moment equivalent to three unpaired electrons ($S=3/2$). A recent extended X-ray-absorphion fine-structure (EXAFS) study[24] summarizes the physical data in support of $(por^{+\cdot})Fe^{IV}{=}O$ as a formulation for Compound I, and $(por)Fe^{IV}{=}O$ for Compound II, and concludes that both species contain an oxene-ferryl group (Fe=O) with a bond length of 1.64 Å.

A recent summary[25] of the activation of $O_2$ by cytochrome P-450 (an iron(III)-heme protein with a axial cysteine thiolate ligand) concludes that the reactive form of this mono-oxygenase also contains an oxene-ferryl group $(RS)(por)Fe^V=O$. The mono-oxygenase chemistry of cytochrome P-450 has been modeled via the use of $(TPP)Fe^{III}Cl$ (TPP=tetraphenylporphyrin) and $(OEP)Fe^{III}Cl$ (OEP=octaethylporphyrin) with peracids,[26,27] iodosobenzene,[26,27] 4-cyano-N,N-dimethylaniline-N-oxide,[28] and hypochlorite[29] to oxygenate model substrates. On the basis of the close parallel with the products from the cytochrome P-450–catalyzed reactions and the net two-oxidizing equivalents of the catalytic cycles for cyt $P-450/(O_2 + 2H^+ + 2e^-)$ and HRP/HOOH, a general consensus has developed that the reactive intermediate of cytochrome P-450 is analogous to Compound I of horseradish peroxidase (HRP-I) with an Fe=O group.

All contemporary work indicates that the reactive intermediate for HRP-I and cytochrome P-450 is an oxygen-atom adduct of $[(imid)^+Fe^{III}(por)]$ and $(por)Fe^{III}-SR$.[24,30] The common belief is that atomic oxygen invariably removes two electrons from iron(III) and/or $Fe(por)^+$ to achieve an oxo $(O^{2-})$ state. Although this misconception is general for the oxygen compounds of transition metals, there is no thermodynamic, electronegativity, or theoretical basis to exclude stable $M-O^-$ and M=O species.[31,32]

The results of recent investigations[33] of model systems provide compelling evidence that stabilized atomic oxygen is present in Compound I and Compound II of horseradish peroxidase. Thus, the combination of tetrakis(2,6-dichlorophenyl)-porphinato-iron(III) perchlorate (6, Scheme 4-4) with pentafluoro-iodosobenzene, *m*-chloroperbenzoic acid, or ozone in acetonitrile at -35°C yields a green porphyrin-oxene adduct (7). This species, which has been characterized by spectroscopic, magnetic, and electrochemical methods, cleanly and stereospecifically epoxidizes olefins (>99% exo-norbornene-oxide).

The reaction chemistry of Scheme 4-4 illustrates how 7 acts as an oxygen-atom-transfer agent towards olefins. The stereospecificity for the epoxidation of norbornene is consistent with the concerted insertion[34] of a singlet oxygen atom into the π bond [analogous to the stereospecific transfer of a singlet oxygen atom from uncatalyzed *m*-ClPhC(O)OOH to norbornene]. If 7 contained hypervalent iron, an electron-transfer mechanism would be favored, which results in a mixture of exo and endo epoxide.[27,35]

The spectroscopy, electrochemistry, and magnetic properties of 7 indicate that its iron center is equivalent to that of Compound I of HRP. Again, the spectroscopic and electrochemical properties of 8, and its reduced reactivity with olefins, indicate that the electronic structure of its iron–oxygen center is analogous to that of Compound II of HRP.

Species 7 contains a stabilized oxygen atom, and the parallel chemistry with the active form of cytochrome P-450 prompts the conclusion that it also contains stabilized atomic oxygen. We have argued elsewhere[20,31] that the most reasonable electronic formulation for the active form of cytochrome P-450 is $(RS)(por)Fe^V=O$ with an RS–Fe covalent bond and an Fe=O covalent double bond.

*Scheme 4-4    Models for Compounds I and II of horseradish peroxidase*

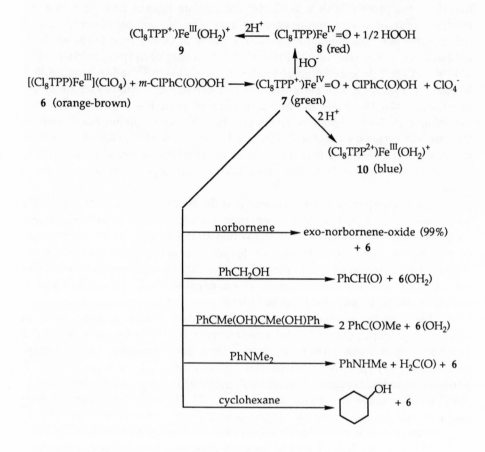

## Activation of HOOH for dioxygenase chemistry

*$Fe^{II}(PA)_2$-induced activation of HOOH*

The addition of HOOH to 2:1 pyridine/acetic acid solutions that contain $Fe^{II}(PA)_2$ (PA = picolinate ion) and cyclohexane ($c$-$C_6H_{12}$) results in the catalyzed transformation of $c$-$C_6H_{12}$ to cyclohexanone [$c$-$C_6H_{10}(O)$].[36] Table 4-5 summarizes the conversion efficiencies and product yields for the oxygenation by the HOOH/Fe(PA)$_2$ combination of several organic substrates (hydrocarbons with methylenic carbons, acetylenes, and aryl olefins). Catalyst turnovers (moles of product per mole of catalyst) are also tabulated. The relative reaction efficiencies for cyclohexane, $n$-hexane, cyclohexene, and 1,4-cyclohexadiene are roughly proportional to the number of ($\supset CH_2$) groups per molecule (6, 4, 4, and 2), and the product for each is the ketone from the transformation of a single methylenic carbon. Addition of a second increment of 56-mM HOOH to a reacted cyclohexane system (Table 4-5) results in an additional ketonization (68% efficiency). The conversion of 1,4-cyclohexadiene to phenol (apparently via ketonization of a methylenic carbon) without any epoxide formation confirms the selectivity of the reactive intermediate. Likewise, the ketonization of cyclohexene further supports the selective reactivity toward methylenic carbon.

The lower reactivity of cyclohexanol relative to cyclohexane (~1/3) indicates that $c$-$C_6H_{11}OH$ is not an intermediate for the ketonization of $c$-$C_6H_{12}$. This is further supported by the results for a combined substrate of 1 $M$ $c$-$C_6H_{12}$ and 1 $M$ $c$-$C_6H_{11}OH$, which has a ketonization efficiency of 65% (in contrast to 72% for 1 $M$ $c$-$C_6H_{12}$ alone, Table 4-5). Likewise, the presence of 1 $M$ $i$-PrOH with 1 $M$ $c$-$C_6H_{12}$ causes a reduction in the conversion efficiency for $c$-$C_6H_{12}$ to 56%, but yields no acetone. Analysis of the product solution during the course of the ketonization of 1 $M$ $c$-$C_6H_{12}$ gives a constant 19:1 $c$-$C_6H_{10}(O)/c$-$C_6H_{11}OH$ ratio (0.1 to 1.0 fractional reaction). With the $Fe^{II}(PA)_2/HOOH/(py_2/HOAc)$ system the reactive intermediate dioxygenates acetylenes to give the α-dione as the sole product; arylolefins are dioxygenated and epoxidized.

The results[36] establish that the pyridine/HOAc (molar ratio, 2:1) solvent system is optimal for the efficient and selective ketonization of methylenic carbons by the $Fe^{II}(PA)_2/HOOH$ system. On the basis of the relative reaction efficiencies for $Fe^{II}(PA)_2$ and $(PA)_2Fe^{III}OFe^{III}(PA)_2$, the initial step when $Fe^{II}(PA)_2$ is used as the catalyst is its transformation to $(PA)_2Fe^{III}OFe^{III}(PA)_2$ **12 b**, Eq. (4-38)]. The spectrophotometric, electrochemical, and magnetic results[37] for the combination of $Fe(PA)_2$ and HOOH in DMF confirm a 2:1 reaction stoichiometry to give a binuclear product ($k_1 = 2 \times 10^3$ $M^{-1}$ $s^{-1}$) [Eq. (4-38)].

$$2\ Fe^{II}(PA)_2 + HOOH \xrightarrow{k_1} \left[ (PA)_2Fe \overset{\overset{\displaystyle H}{\underset{\displaystyle |}{O}}}{\underset{\underset{\displaystyle H}{\underset{\displaystyle |}{O}}}{\diagdown\diagup}} Fe(PA)_2 \longrightarrow (PA)_2Fe^{III}OFe^{III}(PA)_2 + H_2O \right]$$

$$\underset{\textbf{11}}{\phantom{.}} \qquad\qquad\qquad\qquad \underset{\textbf{12a}}{\phantom{.}} \qquad\qquad\qquad \underset{\textbf{12b}}{\phantom{.}}$$

$$(4\text{-}38)$$

Electrochemical measurements[37] establish that (1) auto-oxidation of $Fe^{II}(PA)_2$ in MeCN yields a product that is a mixture of **12a** and **12b** and (2) the product from the 1:1 combination of $Fe^{III}(PA)_3$ and $HO^-$ in DMF is mainly **12b** [Eq. (4-39)].

$$2\ Fe^{III}(PA)_3 + 2\ ^-OH \rightarrow (PA)_2Fe^{III}OFe^{III}(PA)_2 + 2\ PA^- + H_2O \qquad (4\text{-}39)$$
$$\underset{\textbf{12b}}{\phantom{....................}}$$

The addition of species **12a** to excess HOOH in DMF results in the near stoichiometric production of $^1O_2$.[36]

$$\textbf{12a} + HOOH \xrightarrow{\phantom{xx}} [(PA)_2Fe^{III}OOFe^{III}(PA)_2] \longrightarrow {}^1O_2 + 2\ Fe^{II}(PA)_2$$

$$\Big\downarrow \qquad\qquad \underset{\textbf{14}}{\phantom{....}} \qquad\qquad\qquad\qquad \Big\lfloor\underset{\phantom{x}}{\underline{HOOH}}\rightarrow \textbf{12a}$$

$$H_2O$$

$$(4\text{-}40)$$

For the conditions of the experiments that are summarized in Table 4-5 (excess HOOH added to catalyst/substrate), the reaction sequence of Eqs. (4-38) and (4-40) prevails to a major degree (with no evidence of Fenton chemistry in the

Table 4-5 Products and Conversion Efficiencies for the Fe(PA)$_2$-catalyzed (3.5 mM) Ketonization of Methylenic Carbon and the Dioxygenation of Acetylenes and Aryl Olefins by HOOH (56 mM) in Pyridine/HOAc (2:1 mol ratio)[a]

| Substrate (1 M) | Reaction[b] efficiency (%) (±3) | Catalyst[c] turnovers | Products[d] |
|---|---|---|---|
| cyclohexane | 72 | 6 | cyclohexanone (97%), cyclohexanol (3%) |
| n-hexane | 52 | 4 | 3-hexanone (53%), 2-hexanone (46%), 1-hexanol (<2%) |
| PhCH$_2$CH$_3$ | 51 | 5 | PhC(O)CH$_3$ (>96%) |
| PhCH$_2$Ph (0.6 M) | 35 | 3 | PhC(O)Ph (>96%) |
| PhCH$_3$ | 9 | <1 | PhCH(O) (>96%) |
| 2-methyl-butane | 32 | 3 | 3-methyl-2-butanone (>95%), 2-methyl-2 -butanol (<2%) |
| adamantane (0.1 M) | 32 | 3 | 2-adamantanone (43%), 1-adamantanol (29%), 1-pyridyl-adamantane (two isomers, 18% and 10%) |
| cyclododecane (0.5 M) | 70 | 6 | cyclododecanone (90%), cyclododecanol (10%) |
| cyclohexene | 59 | 5 | 2-cyclohexene-1-one (>95%) |
| 1,3-cyclohexadiene | 33 | 5 | PhH (>95%) |
| 1,4-cyclohexadiene | 30 [70][e] | 3 [11] | PhOH (17%), [PhH] (83%) |
| cyclohexanone | 0 | | |
| cyclohexanol | 25 | 4 | cyclohexanone (>95%) |

| | | | |
|---|---|---|---|
| PhC≡CPh (0.6 M) | 40 | 3 | PhC(O)C(O)Ph (>97%) |
| c-PhCH=CHPh | 36 | 4 | PhCH(O) (75%), PhCH–CHPh (25%) (epoxide) |
| t-PhCH=CHMe | 48 | 4 | PhCH(O) (63%), PhCH–CHMe (16%) (epoxide), two others (21%) |

[a] Substrate and $Fe^{II}(TPA)_2$ combined in 3.5 mL of pyridine/HOAc solvent (2:1 mol ratio), followed by the slow addition (1–2 minutes) of 13 μL of 17.3 M HOOH in (49%) $H_2O$ or 60–100 μL of 1.6–3.8 M HOOH (92%) in MeCN to give 56 mM HOOH. Reaction time and temperature; 4 h at $22 \pm 2°C$.

[b] 100% represents one substrate oxygenation per two HOOH molecules added; the remainder of the HOOH was unreacted or consumed via slow $O_2$ evolution and Fenton chemistry to produce $1/n$ [py(OH)]$_n$.

[c] Moles of substrate oxygenated per mole of catalyst.

[d] The product solution was analyzed by capillary gas chromatography and GC-MS (either direct injection of the product solution, or by quenching with water and extracting with diethyl ether).

[e] 100% represents one substrate dehydrogenation per HOOH.

product profiles). The results of Table 4-5 indicate that the relative reactivity of species **14** with hydrocarbon substrates is in the order $>CH_2>PhC\equiv CPh>>$ $ArCH=CHR>>Ar-CH_3>> \geq CH$, which is completely at odds with radical processes.[9]

The results of Table 4-5 together with the data and discussions for the $[Fe^{II}(MeCN)_4](ClO_4)_2/2$ HOOH system[10,11] prompt the formulation of reaction steps and pathways for the $(PA)_2Fe^{III}OFe^{III}(PA)_2/HOOH/(py/HOAc)/substrate$ system (Scheme 4-5). On the basis of the product profiles and reaction efficiencies when $Fe^{II}(PA)_2$ (**11**) is used as the activator, the initial step in the catalytic reaction cycle appears to be the formation of an HOOH adduct $[(PA)_2Fe^{II}(HOOH)$, **13**], which reacts with a second $Fe^{II}(PA)_2$ to give $[(PA)_2Fe^{III}OFe^{III}(PA)_2]$ (**12**). In the presence of $>CH_2$ or $ArCH=CHR$ groups, species **12** rapidly forms (with another HOOH) the activated complex (species **14**, Scheme 4-5). The precatalyst (species **13**) reacts with selective substrates in a manner that is analogous to that of other iron–HOOH adducts.[21,22,24,25] Thus, for conditions that favor formation of species **13** [1:1 $Fe^{II}(PA)_2/HOOH$ in MeCN] mono-oxygenation of hydrocarbons to alcohols dominates, and epoxidation of $c$-PhCH=CHPh is enhanced (via species **15**), but for conditions that favor species **14** [1:20 $Fe^{II}(PA)_2/HOOH$ in $py_2(HOAc)$] ketonization of methylenic carbons and dioxygenation is the dominant path.

Species **14** transforms methylenic carbons $>CH_2$ to ketones ($>C=O$) and dioxygenates acetylenes and aryl olefins, which parallels the reactivity of $[(Ph_3PO)_4(H_2O)Fe^{III}OOFe^{III}(OH_2)(OPPh_3)_4](ClO_4)_4$ (**16**).[16]

$$[(Ph_3PO)_4(H_2O)Fe^{III}OOFe^{III}(OH_2)(OPPh_3)_4](ClO_4)_4 + PhC\equiv CPh \xrightarrow{MeCN}$$
$$\underset{\textbf{16}}{}$$

$$PhC(O)C(O)Ph + 2\ Fe^{II}(OPPh_3)_4^{2+} + 2\ H_2O \qquad (4\text{-}41)$$

Species **16** is formed from the combination of $Fe^{II}(OPPh_3)_4^{2+}$ and HOOH in acetonitrile.[16]

$$2\ Fe^{II}(OPPh_3)_4^{2+} + 2\ HOOH \rightarrow [2\ L_4^{2+}Fe^{IV}=O] + 2\ H_2O$$
$$(\text{or HOIO}_3) \qquad \textbf{17} \mid \qquad (\text{or HOIO}_2)$$
$$\longrightarrow L_4^+(H_2O)^+Fe^{III}OOFe^{III}(OH_2)^+L_4^+$$
$$\underset{\textbf{16}}{}$$
$$(4\text{-}42)$$

The reaction paths for $Fe^{II}(DPA)$ (DPA = 2,6-dicarboxylato-pyridine) and $(DPA)Fe^{III}OFe^{III}(DPA)$, which are the most effective and selective catalysts of the iron complexes investigated, appear to be analogous to those for $Fe^{II}(PA)_2$ (**11**) and $(PA)_2Fe^{III}OFe^{III}(PA)_2$ (**12**) [Eqs. (4-38) and (4-40), and Scheme 4-5].

*Scheme 4-5   Activation of HOOH by $Fe^{II}(PA)_2$ in $py_2(HOAc)$*

*a.  Reaction paths*

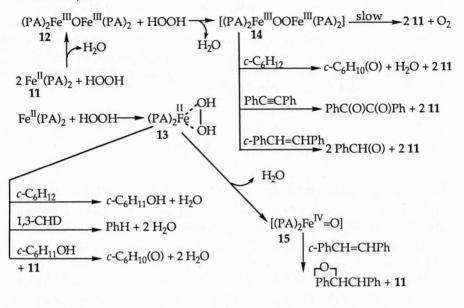

*b.  Proposed mechanisms*

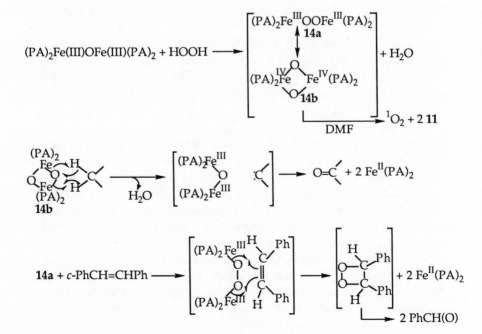

Scheme 4-5b outlines a proposed concerted, dehydrogenation/singlet–singlet mono-oxygenation mechanism for the selective ketonization of methylenic carbons and a direct dioxygenation of arylolefins via a common reactive intermediate, species **14** (resonance hybrids **14b** and **14a**, respectively), which evolves $^1O_2$ in substrate-free DMF.[36] The ketonization of a methylene group in cyclohexene and 1,4-cyclohexadiene (Table 4-5) is especially compelling evidence for a concerted selective process that is optimal for the geometry of methylenic carbons. In contrast, the dehydrogenation of 1,3-cyclohexadiene (Table 4-5) probably is the result of selective reactivity with the precatalyst, species **13** (Scheme 4-5).

## $Co^{II}(bpy)_2^{2+}$-induced activation of HOOH

In acetonitrile/pyridine (4:1 molar ratio) bis(bipyridine)-cobalt(II) $[Co^{II}(bpy)_2^{2+}, \mathbf{18}]$ catalytically activates HOOH via the initial formation of an oxene intermediate $[(bpy)_2^{2+}Co^{III}O\cdot, \mathbf{19}]$ that leads to a dioxygenase $[(bpy)_2^{2+}Co^{III}(OO), \mathbf{20}]$.[38] The latter species selectively ketonizes methylenic carbons and dioxygenates arylolefins, and species **19** epoxidizes olefins. Table 4-6 summarizes the product distributions (for a series of substrates) that result from the catalytic activation of HOOH or $t$-BuOOH by $Co^{II}(bpy)_2^{2+}$ in 4:1 MeCN/py and in pure MeCN. The product profiles indicate that oxidase (or mono-oxygenase) chemistry is favored in pure MeCN solvent ($c$-$C_6H_{12} \rightarrow c$-$C_6H_{11}OH$), but the ketonization of methylenic carbon and dioxygenase chemistry are favored in MeCN/py (4:1 molar ratio) $[c$-$C_6H_{12} \rightarrow c$-$C_6H_{10}(O); c$-$PhCH=CHPh \rightarrow 2\ PhCH(O)]$. The selective ketonization of cyclohexene in MeCN/py contrasts with its enhanced mono-oxygenation in pure MeCN (one/ol ratio; 16:1 versus 1:1) and is compelling evidence for two reactive intermediates. The presence of $O_2$ inhibits the reactivity of $c$-$C_6H_{12}$ with HOOH by 10–20%. In pure MeCN $Co^{II}(bpy)_2^{2+}$ catalyzes HOOH for the stoichiometric transformation of 1,4-cyclohexadiene to benzene.

When $t$-BuOOH is the oxygen source the reactivity with substrates is about ten times greater in pure MeCN than in MeCN/py (Table 4-6). With $PhCH_3$ the dominant product is $PhCH_2OOBu$-$t$, which requires two $t$-BuOOH molecules per substrate. When $c$-$C_6H_{12}$ is the substrate, $c$-$C_6H_{10}(O)$ and $c$-$C_6H_{11}OOBu$-$t$ are the major products (both require two $t$-BuOOH molecules per substrate), and the ketone probably results from the decomposition of $c$-$C_6H_{11}OOBu$-$t$. In contrast, with $(Me)_2CHCH_2Me$ the major product is $(Me)_2C(OH)CH_2Me$ (one $t$-BuOOH per substrate). The use of $t$-BuOOH precludes (or strongly suppresses) formation of the reactive intermediate for the direct ketonization of methylenic carbons.

The results of Table 4-6 and the close parallels of the product profiles to those for the $Fe^{II}(PA)_2/HOOH/(py/HOAc)$ system[36] prompt the conclusion that the combination of $Co^{II}(bpy)_2^{2+}$(**18**) and HOOH results in the initial formation of an oxene intermediate $[(bpy)_2^{2+}Co^{III}O\cdot, \mathbf{19}]$, which (in MeCN/py) rapidly reacts with a second HOOH to give a dioxygenase reactive intermediate $[(bpy)_2^{2+}Co^{III}(O_2), \mathbf{20}]$ (Scheme 4-6).

Table 4-6 Activation of HOOH and $t$-BuOOH by $Co^{II}(bpy)_2^{2+}$ for the Oxygenation of Hydrocarbons, the Oxidation of Alcohols and Aldehydes, and the Dioxygenation of Arylolefins and Acetylenes in 4:1 MeCN/py[a]

| Substrate (1 M) | Oxidant (0.2 M) | Products (mM)[b] |
|---|---|---|
| c-C$_6$H$_{12}$ | HOOH | c-C$_6$H$_{10}$(O) (61), c-C$_6$H$_{11}$OH (1) |
| c-C$_6$H$_{12}$ (MeCN) | HOOH | c-C$_6$H$_{10}$(O) (14), c-C$_6$H$_{11}$OH (9) |
| c-C$_6$H$_{12}$ | t-BuOOH | c-C$_6$H$_{11}$OOBu-t (1.5) |
| c-C$_6$H$_{12}$ (MeCN) | t-BuOOH | c-C$_6$H$_{10}$(O) (15), c-C$_6$H$_{11}$OOBu-t (2), c-C$_6$H$_{11}$OH (1) |
| Me$_2$CHCH$_2$Me | HOOH | Me$_2$CHC(O)Me (12), Me$_2$C(OH)CH$_2$Me (5) |
| Me$_2$CHCH$_2$Me (MeCN) | t-BuOOH | Me$_2$C(OH)CH$_2$Me (9), Me$_2$CHC(O)Me (1) |
| PhCH$_2$CH$_3$ | HOOH | PhC(O)Me (30), PhCH$_2$CH$_2$OH (11) |
| PhCH$_3$ | HOOH | PhCH(O) (20), PhCH$_2$OH (17) |
| PhCH$_3$ (MeCN) | t-BuOOH | PhCH$_2$OOBu-t (28), PhCH(O) (12) |
| c-C$_6$H$_{10}$ | HOOH | R-one (50)[c], epoxide (8), R-OH (3)[d] |
| c-C$_6$H$_{10}$ (MeCN) | HOOH | R-OH (31), R-one (30), epoxide (12), R-R (1) |
| c-C$_6$H$_{10}$ (MeCN) | t-BuOOH | R-OOBu-t (41), R-one (6), R-OH (3), R-R (1) |

| | | |
|---|---|---|
| PhH (MeCN) | HOOH | PhOH (34) |
| c-C$_6$H$_{11}$OH (MeCN) | HOOH | c-C$_6$H$_{10}$(O) (28) |
| PhCH$_2$OH (MeCN) | HOOH | PhCH(O) (40) |
| PhCH(O) (MeCN) | HOOH | PhC(O)OH (108) |
| c-PhCH=CHPh (0.65 M) | HOOH | PhCH(O) (87), epoxide (4) |
| PhC≡CPh | HOOH | PhC(O)C(O)Ph (24) |
| 2,6-(Me)$_2$PhOH | HOOH | 2,6-(Me)$_2$Ph(O)$_2$ (5),[e] ROOR (3) |
| 2,6-(Me)$_2$PhOH (MeCN) | t-BuOOH | ROOR (9) |

[a] Substrates and catalyst combined in 7 mL of MeCN/py (4:1 molar ratio) (or MeCN), followed by the slow addition (1–2 min) of either 100 µL of 17.6 $M$ HOOH (50% in H$_2$O) to give 200 mM HOOH, or 600 µL of 3.0 $M$ t-BuOOH (in 2,2,4-trimethyl-pentane) to give 200 mM t-BuOOH. Reaction time and temperature: 6 h at 22±2°C. [b] The product solutions were analyzed by capillary gas chromatography and GC-MS (either direct injection of the product solution, or by quenching with H$_2$O and extracting with diethyl ether).

[c] c-C$_6$H$_8$-2-ene-1-one.

[d] c-C$_6$H$_8$-2-ene-1-ol.

[e] 2,6-(Me)$_2$-$p$-benzoquinone.

Scheme 4-6  Activation of HOOH and t-BuOOH by $Co^{II}(bpy)_2^{2+}$

a. *HOOH (MeCN/py); [MeCN]*

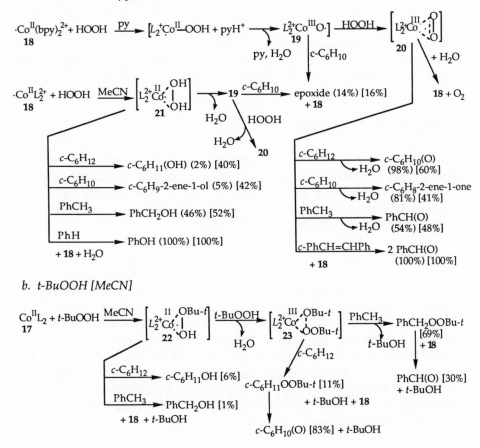

b. *t-BuOOH [MeCN]*

In pure MeCN species **18** appears to activate HOOH and $t$-BuOOH via formation of 1:1 adducts [(bpy)$_2^{2+}$Co$^{II}$(HOOH), **21** and (bpy)$_2^{2+}$Co$^{II}$($t$-BuOOH), **22**], which, when formed in the presence of substrates, act as mono-oxygenases ($c$-$C_6H_{12}$ → $c$-$C_6H_{11}$OH). As such, they are closely similar to the reactive intermediate from the combination of [Fe$^{II}$(MeCN)$_4$](ClO$_4$)$_2$ and HOOH in MeCN.[10,11] The formation of two reactive intermediates [**21**, favored in MeCN, and **20**, favored in MeCN/py] in combination with the product profiles of Table 4-6 is the basis for the proposed reaction pathways of Scheme 4-6. Species **20** transforms methylenic carbons ( $\diagdown$CH$_2$ ) to ketones ( $\diagdown$C=O) and dioxygenates arylolefins and acetylenes, and its precursor (species **19**) epoxidizes aliphatic olefins. Combination of $t$-BuOOH and Co$^{II}$(bpy)$_2^{2+}$ appears to form intermediates **22** and **23**; species **22** has similar reactivity to species **21**, but species **23** is unique and necessary to account for the observed ROOBu-$t$ products.

In summary, the $Co^{II}(bpy)_2^{2+}$/HOOH/(4:1 MeCN/py) system forms a reactive intermediate (20) that selectively ketonizes methylenic carbon, and as such is closely similar to the intermediate of the $Fe^{II}(PA)_2$/HOOH/(2:1 py/HOAc) system.[36] The ability of $Fe^{II}(DPAH)_2$ to activate $O_2$ to an intermediate that has the same unique selectivity for hydrocarbon ketonization[39] is further support for a common stabilized-dioxygen reactive complex (see Chapter 6). Several cobalt-dioxygen complexes exhibit oxygenase reactivity with organic substrates,[40,41] which is consistent with the dioxygen formulation for species 20.


## Nucleophilic character of HOOH

Because the HO–OH bond in hydrogen peroxide is weak ($\Delta H_{DBE}$, 51 versus 90 kcal for the H–OOH bond), there has been a tendency to assume homolytic cleavage in the reactivity and activation of HOOH. However, in a pioneering study of the conversion by HOOH of sulfur dioxide to sulfuric acid Halperin and Taube[42] proposed a nucleophilic-addition mechanism.

$$HO^*O^*H + SO_2 \rightarrow [HO^*O^*S(O)OH] \rightarrow HO^*S(O^*)(O)OH \qquad (4\text{-}43)$$

In a subsequent investigation[43] of this system with anhydrous HOOH in dry MeCN, we discovered that there is no net reaction in the absence of $H_2O$. The reaction is first order with respect $H_2O$ up to a 200-m$M$ concentration, and becomes second and third order at higher concentrations. The reactivity of $t$-BuOOH with $SO_2$ also exhibits a similar first order dependence on $H_2O$ concentration, but $t$-BuOOBu-$t$ is unreactive under all conditions (the $t$-BuO–OBu-$t$ bond energy is essentially the same as that for the HO–OH bond). Thus, water induces HOOH ($pK_a$, 11.8) to act as a nucleophile towards $SO_2$ ($pK_a$, 1.9) with the reaction rate fastest at $pH$ 1.9.

$$HOOH + H_2O \rightleftharpoons H_3O^+ + HOO^-, \quad K,\ 10^{-11.8} \qquad (4\text{-}44)$$

$$HOOH + SO_2 + H_2O \rightarrow [HOOS(O)O^- + H_3O^+]$$

$$\downarrow$$

$$(HO)_2S(O)_2 \cdot H_2O \rightarrow HOS(O)_2O^- + H_3O^+ \qquad (4\text{-}45)$$

If HOOH can act as a nucleophile in aqueous solutions at $pH$ 1.9, then stronger bases in aprotic media [e.g., pyridine in $py_2$(HOAc) and MeCN solvents] should facilitate similar pathways. Hence, the Fenton process in $H_2O$ (at $pH$ 2 to 10)[8] and in aprotic media [$py_2$(HOAc)][9] must involve nucleophilic addition

$$(PA)_2Fe^{II} + HOOH + py \rightarrow [(PA)_2^- Fe^{II}-OOH + Hpy^+]$$

$$[(PA)_2Fe^{III}OH + HO\cdot + py] \qquad (4\text{-}46)$$

Likewise, the activation of HOOH by horseradish peroxidase to Compound I probably involves a similar pathway, for example,

$$(Por)Fe^{III}(imid)^+ + HOOH + imid \longrightarrow \left[ (imid)Fe\overset{III}{O}OH \right] + (imid)H^+$$

$$(imid)Fe\overset{+}{\underset{}{=}}\overset{IV}{O} + H_2O + imid \qquad (4\text{-}47)$$

Compound I

The base-induced decomposition of HOOH almost certainly is another example of nucleophilic addition.[7]

$$HOOH \xrightarrow{\;HO^-\;} HOO^- \xrightarrow{\;HOOH\;} [HOOOH] + HO^- \qquad O_2^- + H_2O$$
$$\searrow H_2O \qquad\qquad\qquad \xrightarrow{\;py\;} \tfrac{1}{n}[pyOH]_n + HOO\cdot \qquad \overset{HO^-}{} \qquad (4\text{-}48)$$

With an effective trap for HO· (such as pyridine), the net product is $(pyOH)_n$ polymer and superoxide ion ($O_2^-\cdot$).

A likely example of HOOH-nucleophilicity is the formation of singlet dioxygen ($^1O_2$) via the combination of hypochlorous acid [HOCl, $pK_{a(H_2O)}$ 7.1] and HOOH.

$$HOOH + HOCl \xrightarrow{\;H_2O\;} [HOOCl] + 2\,H_2O\ (HO^- + H_3O^+) \qquad (4\text{-}49)$$
$$\longrightarrow {}^1O_2$$
$$HCl$$

The proposed intermediate [HOOCl] will yield $^1O_2$ if H-Cl is homolytically formed ($\Delta H_{DBE}$, 103 kcal) via intramolecular bond formation.

From these examples, nucleophilic addition of HOOH to electrophilic centers appears to be general. Thus, the activation of HOOH by $Co^{II}(bpy)_2^{2+}$ and $Fe^{II}(PA)_2$ in the presence of pyridine occurs via the steps of Schemes 4-6 and 4-7.

*Scheme 4-7 Activation via nucleophilic addition of HOOH to Fe$^{II}$(PA)$_2$*

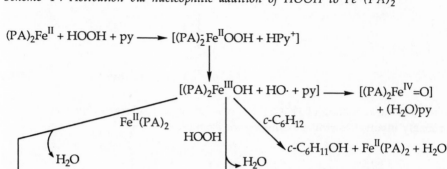

## Biological systems

Aerobic organisms make constructive use of dioxygen via redox chains in respiration and oxidative phosphorylation, and for selective substrate transformations that are catalyzed by enzymes (oxidases and oxygenases). Many oxidases act as dehydrogenases to give HOOH as a co-product. Nature, in most cases, makes use of this HOOH via metalloproteins that facilitate energy "harvesting" and constructive substrate transformations (e.g., peroxidases and myeloperoxidase). However, one protein system is present in all aerobic organisms to facilitate the safe destruction of HOOH via disproportionation to $O_2$ and $H_2O$ (in animals this is a heme protein, catalase; and in plants it is a manganese-protein, pseudo-catalase). Each class of HOOH-activating enzymes includes a variety of proteins and functions, and it is beyond the scope of this monograph to discuss the chemistry of individual HOOH/enzyme/substrate processes. Instead, the chemistry of HOOH and its activation by transition metals (discussed in the earlier sections of this chapter) is brought to bear on a specific example for each class of HOOH-activating proteins.

### Cytochrome-c Peroxidase

Yeasts, via oxidases acting as dehydrogenases (*D*-glucose oxidase), produce substantial fluxes of HOOH. In order to "harvest" the redox energy of this intermediate, these organisms have a heme protein (cytochrome-c peroxidase) to facilitate the electron-transfer oxidation of two cytochrome c-[Fe(II)] molecules

$[E^{o'}_{pH\,7},\ +0.3$ V versus NHE] per HOOH.  Because the uncatalyzed one-electron reduction of HOOH is disfavored and yields hydroxyl radical,

$$HOOH + (H_3O^+)_{pH\,7} \xrightarrow{\;e^-\;} 2\,H_2O + HO\cdot,\ E^{o'}_{pH\,7},\ +0.40 \text{ V versus NHE} \qquad (4\text{-}50)$$

effective energy transduction requires a one-electron catalyst that will store the two oxidizing equivalents of HOOH.

The cytochrome-$c$ peroxidase protein consists of a heme-iron(III) center with an axial histidine ligand that reacts with HOOH to form a ferryl ($Fe^{IV}=O$)-protein cation radical (Compound I), which in turn transfers two electrons from two reduced cytochrome-$c[Fe(II)]$ proteins.[44]  The general expectation of most contemporary discussions is that electron transfer by Compound I is metal centered rather than oxygen centered.  However, consideration of the electron-transfer thermodynamics for oxygen species (Chapter 2) and the bond energetics of oxygen with metals and hydrogen (Chapter 3) prompts a reformulation of the mechanism that is consistent with the chemistry for HOOH/[iron(III)-porphyrins] (see preceding sections of this chapter).  Such a mechanism is proposed and outlined in Scheme 4-8.  In the absence of substrate the reactive intermediate reacts with itself to give Compound I, but in the presence of reduced cytochrome $c$ the $(Fe^{III}–OOH)/H_3O^+$ combination transfers two electrons from two substrate molecules.

*Scheme 4-8  Proposed mechanism for the HOOH/cytochrome-c reaction cycle of cytochrome-c peroxidase*

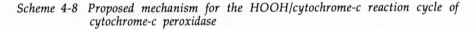

*Peroxidases and catalase*

Horseradish peroxidase, which is the most fully characterized of the heme($Fe^{III}$)-centered enzymes for the activation of HOOH, has an axial histidine ligand; this also is the case for myeloperoxidase (uses HOOH and Cl⁻ to form HOCl).  In contrast, the catalase protein (catalyzes the harmless destruction of HOOH into $O_2$ and $H_2O$) has an axial tyrosine ligand in place of the histidine

*Scheme 4-9    Proposed mechanisms for the activation of HOOH by a*
*(a) horseradish peroxidase, (b) myeloperoxidase, and (c) catalase*

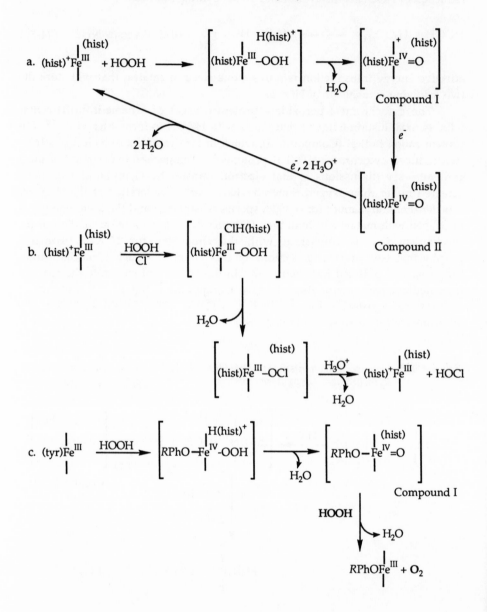

residue. The horseradish peroxidase protein activates HOOH via formation of Compound I [(por$^{+\cdot}$)Fe$^{IV}$=O, Eq. (4-47)], which is (1) an effective one-electron oxidant to give Compound II [(por)Fe$^{IV}$=O], (2) an oxidase/dehydrogenase to give (por)Fe$^{III}$, and (3) a mono-oxygenase to give (por)Fe$^{III}$.[23,24] In the absence of substrate myeloperoxidase and catalase also form a species that is designated as Compound I [two oxidizing equivalents above the (por)Fe(III) state].

Although most biologists consider Compound I of horseradish peroxidase to be a powerful one-electron oxidant by virtue of an electrophilic, hypervalent iron center, the chemistry in the preceding sections of this chapter for various model systems is consistent with a porphyrin cation radical as the high-potential one-electron oxidant, and weakly bonded atomic oxygen as the oxidase/dehydrogenase and mono-oxygenase.[33] On the basis of these considerations mechansims are proposed and outlined for the activation of HOOH by (1) horseradish peroxidase, (2) myeloperoxidase, and (3) catalase in Scheme 4-9. In the absence of chloride ion, myeloperoxidase is transformed by HOOH into a Compound I that appears to be identical to that of horseradish peroxidase.[19]

The presence of the phenolate ligand in catalase gives an uncharged Compound I without a high-potential porphyrin cation radical, and an entity that is especially effective for the concerted removal of the two hydrogen atoms of HOOH to give $H_2O$ and $O_2$ (-$\Delta G_{reaction}$, ~20 kcal; see Chapter 3).

### References

1. Lide, D. R. (ed.). *CRC Handbook of Chemistry and Physics*, 71st ed. Boca Raton, FL; CRC, 1990, pp. 9-86–98.
2. Sawyer, D. T. *Comments Inorg. Chem.* **1990**, *10*, 129.
3. Sugimoto, H.; Tung, H.-C.; Sawyer, D. T. *J. Am. Chem. Soc.* **1988**, *110*, 2465.
4. Cofré, P.; Sawyer, D. T. *Inorg. Chem.* **1986**, *25*, 2089.
5. Calderwood, T. S.; Johlman, C. L.; Roberts, J. L., Jr., Wilkins, C. L.; Sawyer, D. T. *J. Am. Chem. Soc.* **1984**, *106*, 4683.
6. Goolsby, A. D.; Sawyer, D. T. *Anal. Chem.* **1968**, *40*, 83.
7. Roberts, J. L., Jr.; Morrison, M. M.; Sawyer, D. T. *J. Am. Chem. Soc.* **1978**, *100*, 329.
8. Walling, C. *Acc. Chem. Res.* **1975**, *8*, 125.

9.  Sheu, C.; Sobkowiak, A.; Zhang, L.; Ozbalik, N.; Barton, D. H. R.; Sawyer, D. T.;
    *J. Am. Chem. Soc.* **1989**, *111*, 8030.
10. Sugimoto, H., Sawyer, D. T. *J. Am. Chem. Soc.* **1985**, *107*, 5712.
11. Sugimoto, H., Sawyer, D. T. *J. Am. Chem. Soc.* **1984**, *106*, 4283.
12. Sawyer, D. T.; Roberts, J. L., Jr.; Calderwood, T. S.; Sugimoto, H.; McDowell, M. S.
    *Phil. Trans. Roy. Soc. Lond.* **1985**, *B311*, 483.
13. Nicholls, P. *Biochim. Biophys. Acta* **1972**, *279*, 306.
14. Guengerich, F. P.; MacDonald, T. L. *Acc. Chem. Res.* **1984**, *17*, 9.
15. Groves, J. T.; McClusky, G. A.; White, R. E.; Coon, M. J. *Biochem. Biophys. Res.
    Commun.* **1978**, *81*, 154.
16. Sawyer, D. T.; McDowell, M. S.; Spencer, L.; Tsang, P. K. S. *Inorg. Chem.* **1989**, *28*,
    1166.
17. Sugimoto, H.; Sawyer, D. T. *J. Org. Chem.* **1985**, *50*, 1785.
18. Rosen, H.; Klebanoff, S. J. *J. Biol. Chem.* **1977**, *252*, 4803.
19. Held, A. M.; Hurst, J. K. *Biochim. Biophys. Res. Commun.* **1978**, *81*, 878.
20. Sugimoto, H.; Spencer, L.; Sawyer, D. T. *Proc. Natl. Acad. Sci USA* **1987**, 23, 1731.
21. VanAtta, R. B.; Franklin, C. C.; Valentine, J. S. *Inorg. Chem.* **1984**, *23*, 4121.
22. Groves, J. T.; McClusky, G. A. *J. Am. Chem. Soc.* **1976**, *98*, 859.
23. (a) George, P. *Adv. Catal.* **1952**, *4*, 367; (b) George, P. *Biochim. J.* **1953**, *54*, 267;
    (c) George, P. *Biochim. J.* **1953**, *55*, 220.
24. Penner-Hahn, J. E.; Eble, K. E.; McMurry, T. J.; Renner, M.; Balch, A. L.; Groves, J. T.;
    Dawson, J. H.; Hodgson, K. O. *J. Am. Chem. Soc.* **1986**, *108*, 7819.
25. Guengerich, P. F.; McDonald T. L. *Acc. Chem. Res.* **1984**, *17*, 9.
26. (a) Groves J. T.; Haushalter, R. C.; Nakamura, M.; Nemo, T. E.; Evans, B. J. *J. Am.
    Chem. Soc.* **1981**, *103*, 2884; (b) Groves, J. T.; Watanabe, Y. *J. Am. Chem. Soc.*
    **1986**, *108*, 7834.
27. (a) Traylor, P. S.; Dolphin, D.; Traylor, T. G. *J. Chem. Soc. Chem. Commun.* **1984**, 279;
    (b) Traylor, T. G.; Nakano, T.; Dunlap, B. E.; Traylor, P. S.; Dolphin D. *J. Am. Chem.
    Soc.* **1986**, *108*, 2782.
28. (a) Dicken, C. M.; Woon, T. C.; Bruice, T. C. *J. Am. Chem. Soc.* **1986**, *108*, 1636;
    (b) Calderwood, T. S.; Bruice, T. C. *Inorg. Chem.* **1986**, *25*, 3722; (c) Calderwood, T .S.;
    Lee, W. A.; Bruice, T. C. *J. Am. Chem. Soc.* **1985**, *107*, 8272.
29. Collman, J. P.; Kodadek, T.; Brauman, J. I. *J. Am. Chem. Soc.* **1986**, *108*, 2588.
30. Ortiz, de Montellano, P. R. (ed.). *Cytochrome P-450*. New York: Plenum Press, 1986.
31. Sawyer, D. T. *Comments Inorg. Chem. VI*, **1987**, 103.
32. Goddard, W. A., III; Olafson, B. D. *Proc. Nat. Acad. Sci., USA* **1975**, *72*, 2335.
33. Sugimoto, H.; Tung, H.-C.; Sawyer, D. T. *J. Am. Chem. Soc.* **1988**, *110*, 2465.
34. Chance, M.; Powers, L.; Poulos, T.; Chance, B. *Biochemistry*, **1986**, *25*, 1266.
35. Samsel, E. G.; Srinivasan, K.; Kochi, J. K. *J. Am. Chem. Soc.* **1985**, *107*, 7606.
36. Sheu, C.; Richert, S. A.; Cofré, P.; Ross, B., Jr.; Sobkowiak, A.; Sawyer, D. T.; Kanofsky,
    J. R. *J. Am. Chem. Soc.* **1990**, *112*, 1936.
37. Cofré, P.; Richert, S. A.; Sobkowiak, A.; Sawyer, D. T. *Inorg. Chem.* **1990**, *29*, 2645.
38. Tung, H.-C.; Sawyer, D. T. *J. Am. Chem. Soc.* **1990**, *112*, 8214.
39. Sheu, C.; Sobkowiak, A.; Jeon, S.; Sawyer, D. T. *J. Am. Chem. Soc.* **1990**, *112*, 879.
40. Nishinaga, A.; Tomita, H. *J. Mol. Catal.* **1980**, *7*, 179.
41. Matsuura, T. *Tetrahedron* **1977**, *33*, 2869.

42. Halperin, J.; Taube, H. *J. Am. Chem. Soc.* **1952**, *74*, 380.
43. Sobkowiak, A.; Sawyer, D. T. *J. Am. Chem. Soc.* **1991**, *113*, 0000.
44. Anni, H.; Yonetani, T. In *Oxidase and Related Redox Systems* (King, T. E.; Mason, H. S.; Morrison, M., eds.). New York: Alan R. Liss, Inc., 1988, pp. 437–449.

# 5

# REACTIVITY OF OXYGEN RADICALS
## [HO·, RO·, ·O·, HOO·, ROO·, AND RC(O)O·]

*To become a water molecule causes HO· to be the ultimate oxy radical*

Oxygen radicals are defined as those molecules that contain an oxygen atom with an unpaired, nonbonding electron (e.g., HO·). Although triplet dioxygen ($\cdot O_2 \cdot$) and superoxide ion ($O_2^{-}\cdot$) come under this definition, their nonradical chemistry dominates their reactivity, which is discussed in Chapters 6 ($\cdot O_2 \cdot$) and 7 ($O_2^{-}\cdot$). The hydroxyl radical (HO·) is the most reactive member of the family of oxygen radicals [HO·, *RO·*, ·O·, HOO·, *ROO·*, and *RC(O)O·*], and is the focus of most oxygen radical research. In the gas phase the dramatic example of oxygen radical reactivity with hydrocarbon substrates is combustion, which is initiated by HO· (or *RO·* or *MO·*) and propagated by $\cdot O_2 \cdot$ and ·O·.

$$CH_4 + HO\cdot \longrightarrow HOH + \cdot CH_3 \xrightarrow{\cdot O_2\cdot} CH_3OO\cdot \qquad (5\text{-}1)$$
$$\xrightarrow{\cdot O\cdot} CH_3O\cdot$$

These initial steps are highly exothermic, which accelerates the auto-oxidation cycle to an explosive rate to give CO, $CO_2$, and $H_2O$ as the stable products. Within biological matrices the auto-oxidation and peroxidation of lipids and fats from foodstuffs are important examples of oxygen radical chemistry. In general the initiator is HO· (or HOO·) and the auto-oxidation cycle is carried by $\cdot O_2 \cdot$.

$$RCH{=}CHCH_2(CH_2)_nC(O)OR\,' + HO\cdot \longrightarrow RCH{=}CH{-}\overset{\cdot}{C}H(CH_2)_nC(O)OR' \qquad (5\text{-}2)$$

**1**                **2·**

$$\downarrow \cdot O_2\cdot$$

$$RCH{=}CHCH(OO\cdot)(CH_2)_2C(O)OR'$$

$$\downarrow 1$$

$$2\cdot + RCH{=}CHCH(OOH)(CH_2)_nC(O)OR'$$

Both combustion and lipid auto-oxidation are initiated via the creation of a carbon radical (abstraction of a hydrogen atom by HO· from a C–H bond). The rate is inversely proportional to the C–H bond energy. The C–H bond energies of aromatic carbons are so large that the reaction of HO· with benzene usually goes via addition. Thus, conjugated systems can stabilize radicals by delocalization throughout the $\pi$-electron manifold. However, with saturated $\sigma$-bonded substrates the only pathway for oxygen radical reactions is by hydrogen-atom abstraction.

$$R-H + YO\cdot \longrightarrow YO-H + R\cdot \qquad (5\text{-}3)$$

The driving force for such a reaction is the difference between the free energy of bond formation for YO–H (-$\Delta G_{BF}$), and the enthalpy of bond dissociation for R–H ($\Delta H_{DBE}$). Table 5-1 summarizes the YO–H bond-formation energies for oxygen radicals (YO·), as well as dissociative bond energies for several R–H substrates.[1,2] Reference to these data indicates that $t$-BuO· will react exothermally with toluene (PhCH$_3$), but will be essentially unreactive with methane.

Table 5-1  Radical Strength of Oxygen Radicals (YO·) in Terms of Their YO–H Bond-Formation Free Energies (-$\Delta G_{BF}$), and Dissociative Bond Energies ($\Delta H_{DBE}$) for H–R  Molecules

| Oxy radical (YO·) | Bond (YO–H) | (-$\Delta G_{BF}$)(aq),[a] (kcal mol$^{-1}$) | Bond (H–R) | ($\Delta H_{DBE}$)(g),[b] (kcal mol$^{-1}$) |
|---|---|---|---|---|
| HO· | HO–H | 111 | H–CH$_3$ | 105 |
| O$^-$· | $^-$O–H | 109 | H–($n$-C$_3$H$_7$) | 100 |
| ·O· | ·O–H | 98 | H–($c$-C$_6$H$_{11}$) | 95 |
| $t$-BuO· | $t$-BuO–H | 97 | H–($t$-C$_4$H$_9$) | 93 |
| MeO· | MeO–H | 96 | H–CH$_2$Ph | 88 |
| PhO· | PhO–H | 79 | H–($c$-C$_6$H$_7$)(CHD) | 73 |
| HOO· | HOO–H | 82 | H–C(O)Ph | 87 |
| O$_2^-$· | $^-$OO–H | 72 | H–Ph | 111 |
| ·O$_2$· | ·OO–H | 51 | H–SH | 91 |
| $t$-BuOO· | $t$-BuOO–H | 83 | H–SMe | 89 |
| MeOO· | MeOO–H | 82 | H–SPh | 83 |
| MeC(O)O· | MeC(O)O–H | 98 | H–OPh | 86 |

[a] Ref. 1.
[b] Ref. 2.

### Reactivity of HO·

Gas-phase HO· is generated by photolysis of HOOH and by cosmic radiation and solar radiation of O$_2$/H$_2$O in the atmosphere. The latter process is an important contributor to the atmospheric chemistry associated with organic pollutants. The gas-phase reactions of HO· with organic molecules has been exhaustively

reviewed and summarized in a recent compendium.[3] In aqueous solutions HO·
is produced by radiolysis (continuous and pulsed),[4]

$$HOH \xrightarrow{h\nu} H· + HO·$$
$$\phantom{HOH \xrightarrow{h\nu}}\underset{N_2O}{\big\rfloor} \longrightarrow N_2 + HO·$$                                                                                   (5-4)

and has been the basis for a massive compilation of reactivity data for HO· with
organic and inorganic substrates.[5] Although sophisticated measuring techniques
provide precise kinetic data for the primary (HO·/*R*–H) reaction with pulse-
radiolysis-generated hydroxyl radicals, the limited yields of terminal products
preclude identification and assay. For this reason much of the characterization of
HO· chemistry with organic substrates has made use of the Fenton process.

Fenton chemistry is the use of reduced transition-metal ions [Fe(II),
Mn(II), V(III), and Ti(III)] in stoichiometric combination with HOOH to generate
HO· in situ at a slow rate.[6,7]

$$Fe^{II}(OH_2)_6^{2+} + HOOH + H_2O \xrightarrow[pH\ 2]{k} [(H_2O)_5^{\ddagger} Fe^{II}OOH + H_3O^+], \; k,\, 42\, M^{-1}\, s^{-1} \quad (5\text{-}5)$$

$$\downarrow$$

$$(H_2O)_5^{2+}Fe^{III}OH + HO·$$

With stronger ligands the rate of the Fenton process is enhanced [e.g., Fe$^{II}$(NTA),
$3 \times 10^4\, M^{-1}\, s^{-1}$ (Ref. 8) and Fe$^{II}$(PA)$_2$ in 2:1 py/HOAc, $2 \times 10^3\, M^{-1}\, s^{-1}$ (Ref. 9)]. The
dynamics of (HO·) reactivity with organic substrates are assessed as relative rates
vis-a-vis the fate of Fenton-generated hydroxyl radical in these competitive
reactions.[6]

$$Fe^{II}L_n + HO· \longrightarrow L_nFe^{III}OH, \; k,\, 2 \times 10^8\, M^{-1}\, s^{-1} \qquad (5\text{-}6)$$

$$R\text{–H} + HO· \longrightarrow R· + HOH, \; k,\, 10^8\text{–}10^{10}\, M^{-1}\, s^{-1} \qquad (5\text{-}7)$$

$$HOOH + HO· \longrightarrow HOO· + HOH, \; k,\, 1 \times 10^7\, M^{-1}\, s^{-1} \qquad (5\text{-}8)$$

With 1:1 Fe$^{II}$L$_n$/HOOH at modest concentrations in the presence of excess
substrate, the attack of substrate is favored [Eq. (5-7)].

Although H-atom abstraction is the common path for most substrates, the
large bond energy of aromatic C–H bonds ($\Delta H_{DBE}$, 111 kcal; Table 5-1) precludes
this as a facile process. However, addition to the conjugated $\pi$-electron system is
energetically favored.

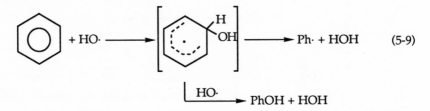

The major limitation of aqueous Fenton generation of HO· is the insolubility of most organic substrates, especially hydrocarbons. The reasonable expectation that those organic solvents that are able to dissolve reasonable quantities of iron complexes are substrates for HO· has precluded their use. However, a recent study[9] has established that the Fenton process [Eq. (5-5)] occurs in pyridine/acetic acid (2:1 molar ratio) when bis(picolinato)iron(II) [$Fe^{II}(PA)_2$] and HOOH are combined in a 1:1 molar ratio

$$Fe^{II}(PA)_2 + HOOH \longrightarrow (PA)_2Fe^{III}OH + HO\cdot, \ k, 2 \times 10^3 \ M^{-1}s^{-1} \quad (5\text{-}10)$$

The cleanness of this system with respect to (1) the generation of HO· radicals and (2) their primary reactivity with hydrocarbon substrates ($RH$, rather than solvent or ligand) has been demonstrated through the use of PhSeSePh as an efficient carbon radical trap.

$$R\text{-}H + HO\cdot \longrightarrow R\cdot + HOH, \ k, \sim10^9 \ M^{-1}s^{-1} \quad (5\text{-}11)$$

$$2\ R\cdot + PhSeSePh \longrightarrow 2\ PhSe\text{-}R \quad (5\text{-}12)$$

Table 5-2 summarizes the reaction efficiencies and product profiles for cyclohexane with the $Fe(PA)_2$/HOOH system in the presence and absence of PhSeSePh for various solvents [the py/HOAc (1.8:1 mol-ratio matrix is optimal]. The product profiles for four other hydrocarbon substrates with the relative abundances of the PhSe–$R$ isomers are included in Table 5-2.

The results for the $Fe^{II}(PA)_2$/HOOH combination (1:1 mol ratio) with cyclohexane in the absence of PhSeSePh (Table 5-2) confirm that the primary step is Fenton chemistry[6] to produce one HO· per HOOH [Eq. (5-10)]. The production of ($c$-$C_6H_{11}$)–py as the major product as well as the formation of significant amounts of ($c$-$C_6H_{11}$)$_2$ (Table 5-2) are consistent with HO· radical chemistry [Eq. (5-11).[5,6]

$$HO\cdot + py \rightleftharpoons [(p\dot{y}(OH)] \longrightarrow (1/n)\,[py(OH)]_n, \ k, 3 \times 10^9 \ M^{-1}s^{-1} \quad (5\text{-}13)$$

Although carbon radicals ($R\cdot$) are trapped by pyridine to give $R$–py,[11] the presence of substantial fluxes of [(p$\dot{y}$(OH)] should favor radical–radical coupling (Scheme 5-1). When the HOOH/Fe(II)(PA)$_2$ ratio is large, the $Fe^{II}(PA)_2$ (at low concentration) is rapidly transformed to $(PA)_2Fe^{III}OFe^{III}(PA)_2$, which activates the remaining HOOH for the ketonization of methylenic carbons (Table 5-2 and Scheme 5-1, and Chapter 4)[12] and eliminates reduced iron for the Fenton process.

Table 5-2  Phenylselenization of Hydrocarbons by a Fenton-Chemistry [Fe$^{II}$(PA)$_2$ + (HOOH)]/PhSeSePh System[a]

a.  Cyclohexane (1 M); 19 mM Fe$^{II}$(PA)$_2$, 19 mM HOOH, 10 mM PhSeSePh

| Conditions | Reactn efficiency (%) (±4) | Products (%(±4)[c] | | | | | |
| --- | --- | --- | --- | --- | --- | --- | --- |
| | | PhSe–(c-C$_6$H$_{11}$) | PhSe–py | c-C$_6$H$_{10}$(O) | c-C$_6$H$_{11}$OH | (c-C$_6$H$_{11}$)–py | (c-C$_6$H$_{11}$)$_2$ |
| py/HOAc (mol ratio, 1.8:1) | 100[d] | 93 | 4 | 3 | 0 | 0 | 0 |
| Controls (no PhSeSePH): [Fe(PA)$_2$]  [HOOH] | | | | | | | |
| 19 mM  19 mM | 79 | 0 | 0 | 14 | 0 | 77 | 9 |
| 9 mM   9 mM | 85 | 0 | 0 | 34 | 0 | 63 | 3 |
| 3.3 mM  56 mM | 72 | 0 | 0 | 94 | 6 | 0 | 0 |
| py | 73 | 93 | 6 | 1 | 0 | 0 | 0 |
| py/HOAc (mol ratio, 1:1) | 89 | 92 | 5 | 3 | 0 | 0 | 0 |
| DMF | 20[e] | 0 | 0 | 0 | 1 | 0 | 0 |
| MeCN [100 mM c-C$_6$H$_{12}$, 10 mM Fe(PA)$_2$, 10 mM HOOH, and 5 mM PhSeSePh] | 7 | 57 | 0 | 0 | 43 | 0 | 0 |
| Control (no PhSeSePh and 500 mM c-C$_6$H$_{12}$): | 51 | 0 | 0 | 33 | 65 | 0 | 2 |

b. Various hydrocarbons (RH); 19 mM $Fe(PA)_2$, 19 mM HOOH, 10 mM PhSeSePh; py/HOAc (1.8:1)

| RH (1 M) | Reactn efficiency[b] (%) (±4) | Products (%)(±2)[c] | | |
|---|---|---|---|---|
| | | PhSe-R [relative isomer abundance] (theor.) | PhSe-py | R(O) |
| n-hexane | 76 | 90 | [12:43:45] [6/4/4][f] | 4 | 6 |
| 2-Me-butane | 57 | 93 | [17:8:35:40] (6/3/2/1)[g] | 5 | 2 |
| adamantane (0.1 M) | 40[h] | 69 | [63:37] (12/4) | 26 | 1 |
| PhCH₂CH₃ | 48[i] | 48 | [32:68] (3/2) | 5 | 14 |

[a] Substrate, $Fe^{II}(PA)_2$, and PhSeSePh combined in 3.5 mL of py/HOAc solvent (unless otherwise indicated), followed by the slow addition (1–2 min) of 2–13 μL of 17.3 M HOOH (49%) in $H_2O$ to give 9–56 mM HOOH. Reaction time and temperature; 10 min at 22±2 °C.

[b] 100% represents one substrate oxidation per HOOH [except for production of R(O), R–R, and R–py, which require 2 HOOH]. Remainder of HOOH was consumed

to produce $PhSe(O)OH$, $1/n [pyOH]_n$, or $O_2$.

[c] The product solutions were analyzed by capillary-GC and GC-MS (direct injection of product solution or ether extract).

[d] Addition of more HOOH to the product solution (after consumption of all PhSeSePh) only produced more $c$-$C_6H_{10}(O)$.

[e] Major product, (DMF)–SePh (two isomers, 79% and 20%).

[f] Isomer order; —CH₂SePh and ⟩CHSePh.

[g] Isomer order; —CH₂SePh(2), —CH₂SePh(1), ⟩CHSePh(1), ⟩CHSePh(1) and ⟩CSePh(1). Authentic samples of PhSe–R were prepared from bromides (RBr) or mesylates, or by reaction of R· with PhSeSePh.

[h] About 4% of the product was py–adamantane.

[i] About 33% of the product was R–R (5 isomers).

*Scheme 5-1   Fenton chemistry in py₂(HOAc)*

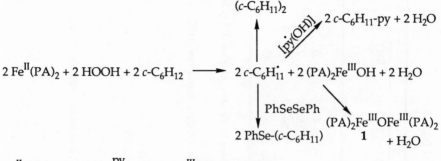

$$Fe^{II}(PA)_2 + HOOH \xrightarrow{py} (PA)_2Fe^{III}OH + [\dot{p}y(OH)]$$

$$2\,Fe^{II}(PA)_2 + HOOH \longrightarrow (PA)_2Fe^{III}OFe^{III}PA)_2$$
$$(< 5\ mM)\qquad (>50\ mM)\qquad\qquad \mathbf{1}$$

In the presence of PhSeSePh and excess hydrocarbon substrate the Fenton process [Fe(PA)₂/HOOH] produces carbon radicals ($R\cdot$), which are trapped by PhSeSePh to give PhSe–$R$ products (Table 5-2 and Scheme 5-1). The distribution of PhSe–$R$ isomers appears to reflect the isomer abundance for the $R\cdot$ radicals from the Fenton cycle [Eqs. (5-10) and (5-11)]. For $n$-hexane and 2-Me-butane the $R$–SePh isomer distribution (Table 5-2) indicates that the relative reaction probabilities of HO· with a C–H bond in $-CH_3$, $\geq CH_2$, and $\geq CH$ groups are 0.074, 0.44, and 1.00 (the respective C–H bond energies are 100, 95, and 93 kcal),[10] which are in accord with the relative values for aqueous HO· (0.10/0.48/1.00).[13] Thus, PhSeSePh provides the means to trap first-formed carbon radicals and thereby give insight to the mechanism of their generation.

With 1:1 Fe$^{II}$(PA)₂/HOOH Fenton generation of HO· is the dominant process, but when the mole ratio of Fe$^{II}$(PA)₂/HOOH is 1:10 or less, the major part of the chemistry does not involve oxygen radicals or reduced iron (Table 5-2a).[12]

The reactivity of RO· (prepared via the Fenton process)

$$ROOH + Fe^{II}L_n \longrightarrow L_nFe^{III}OH + RO\cdot \qquad\qquad (5\text{-}14)$$

is parallel to that of HO·, but less robust because of the smaller bond energy for RO–H (-$\Delta G_{BF}$, 96 kcal) relative to that for HO–H (-$\Delta G_{BF}$, 111 kcal).

## Reactivity of HOO·

The hydroperoxyl radical (HOO·) is the conjugate acid of superoxide ion ($O_2^{-\cdot}$),

$$HOO\cdot + H_2O \rightleftharpoons H_3O^+ + O_2^{-\cdot}, \quad pK_a, 4.9 \qquad (5\text{-}15)$$

and constitutes about 1% of the $O_2^{-\cdot}$ that is formed in aqueous systems at $pH$ 7.[14] Although the O–O bond of HOO· traditionally is viewed to be the same as the single σ bond of HO–OH ($\Delta H_{DBE}$, 51 kcal), its bond energy ($\Delta H_{DBE}$) is about 85 kcal, which is more consistent with the 1.5 bond order of $O_2^{-\cdot}$. However, HOO· is unstable in protic media (such as water and alcohols) and rapidly decomposes via heterolytic and homolytic disproportionation.[15]

$$HOO\cdot + O_2^{-\cdot} \xrightarrow{\;H_2O\;} HOOH + HO^- + O_2, \; k, 1 \times 10^8 \, M^{-1} \, s^{-1} \qquad (5\text{-}16)$$

$$HOO\cdot + HOO\cdot \longrightarrow HOOH + O_2, \; k, 8.3 \times 10^5 \, M^{-1} s^{-1} \qquad (5\text{-}17)$$

Another route to the formation of $HOO\cdot/O_2^{-\cdot}$ is uv irradiation of HOOH in aqueous solutions.[16]

Aerobic organisms produce minor fluxes of superoxide ion ($O_2^{-\cdot}$, and thereby HOO·) during respiration and oxidative metabolism; for example, possibly up to 10–15% of the $O_2$ reduced by cytochrome-$c$ oxidase and by xanthine oxidase passes through the $HOO\cdot/O_2^{-\cdot}$ state.[17] Thus, the chemistry of HOO· (and of $O_2^{-\cdot}$) may be important to an understanding of oxygen toxicity in biological systems.

In aprotic media the rate of protonation of $O_2^{-\cdot}$ is proportional to the acidity of the associated Brønsted acids[18]

$$O_2^{-\cdot} + HA \xrightarrow{\;k\;} HOO\cdot + A^-, \; k \propto K_{HA} \qquad (5\text{-}18)$$

Electrolytic oxidation of HOOH in acetonitrile yields stochiometric fluxes of HOO·.[19]

$$HOOH \xrightarrow[\text{MeCN}]{H_2O} HOO\cdot + H_3O^+ + e^- \qquad (5\text{-}19)$$
$$\qquad\qquad \underset{\text{MeCN}}{\overset{HOO\cdot}{\big\downarrow}} HOOH + O_2, \; k, 1 \times 10^7 \, M^{-1} s^{-1}$$

Reduction of $O_2$ in MeCN at a platinum electrode in the presence of excess protons yields adsorbed Pt(HOO·),[20]

$$2\,O_2 + 2\,H_3O^+ + 2\,Pt + 2\,e^- \longrightarrow \begin{array}{c} Pt\!-\!HOO\cdot \\ | \\ Pt\!-\!HOO\cdot \end{array}$$

$$\begin{array}{c} \downarrow \end{array}$$

$$\begin{array}{c} Pt\!-\!HO \\ | \quad \searrow O \\ Pt\!-\!HO \nearrow O \end{array} \longrightarrow {}^1O_2 + HOOH \qquad (5\text{-}20)$$

which forms [HOOOOH] via radical coupling at the surface and homolytically dissociates to $^1O_2$ and HOOH.

The homogeneous disproportionation of HOO· in aprotic solvents does not yield $^1O_2$ and appears to involve a "head-to-tail" dimer intermediate that undergoes H-atom transfer[20,21]

$$2\,HOO· \longrightarrow [HOO·\ HOO·] \longrightarrow HOOH + ·O_2^-, \ k_d \qquad (5\text{-}21)$$

Table 5-3 summarizes the values of the rate constants ($k_d$) for the homogeneous disproportionation of HOO· in various solvent systems.[14,18,19,21,22] This second-order parameter limits the steady-state concentration (flux) of HOO· in a given solvent and thereby its reactivity with substrates. Hence, dimethyl sulfoxide provides a matrix that stabilizes HOO· ($k_d$, $1.7 \times 10^4\ M^{-1}\ s^{-1}$) better than the other solvents.

Table 5-3   Rate Constants for the Second-Order Disproportionation ($k_d$) of HOO· in Various Solvents at 25°C

| Solvent | $k_d\ (M^{-1}\ s^{-1})$ | Ref. |
|---|---|---|
| Me$_2$SO | $(1.7\pm0.5) \times 10^4$ | 21 |
| DMF | $(5.3\pm0.5) \times 10^4$ | 21 |
| MeCN | $(1.0\pm0.5) \times 10^7$ | 19 |
|  | $>5 \times 10^6$ | 21 |
|  | $(5\pm1) \times 10^6$ | 18 |
|  | $4.3 \times 10^6$ (30°C) | 22 |
| PhCl | $6.3 \times 10^8$ | 22 |
| H$_2$O | $(8.3\pm0.7) \times 10^5$ | 14 |

An early study[22] used radical-initiated auto-oxidation experiments in acetonitrile (MeCN) and in chlorobenzene (PhCl) to demonstrate that HOO· abstracts hydrogen atoms from allylic hydrocarbons [e.g., 1,4-cyclohexadiene (1,4-CHD)]. Subsequent investigations[21] have used $O_2^-$· in Me$_2$SO with limiting fluxes of protons to generate HOO· at a controlled rate [Eq. (5-18)]. Because the disproportionation of HOO· is a second-order process, such limiting conditions favor H-atom abstraction by HOO· from excess 1,4-CHD (especially true for Me$_2$SO solvent in which HOO· disproportionation is slower than in other solvents; Table 5-3). Hence, the presence of 1,4-CHD (1,4-$c$-C$_6$H$_8$) enhances the rate of disappearance of HOO· because of its parallel oxidation of an allylic hydrogen to give a $1,3\text{-}\dot{C}_6H_7$ radical that disproportionates to 1,3-CHD and benzene.

$$HOO\cdot + 1,4\text{-}c\text{-}C_6H_8 \longrightarrow 1,3\text{-}\dot{C}_6H_7 + HOOH, \; k_{ox}, 1.6 \times 10^2 \, M^{-1} \, s^{-1} \quad (5\text{-}22)$$

Analysis of the kinetic data for the decay of the HOO· concentration in such systems provides evaluations of $k_{ox}$.

Table 5-4 summarizes values of $k_{ox}$ in three solvents (Me$_2$SO, MeCN, and PhCl) for the oxidation of 1,4-CHD by HOO·,[21,22] as well as for several other substrates in water/alcohol solvents.[21-26] A reasonable reaction sequence for the formation and reaction of HOO· is outlined in Scheme 5-2. Initial formation of HOO· (at diffusion-controlled rates for strong acids) is followed by its disproportionation $(k_d)$ or attack of 1,4-CHD $(k_{ox})$. As with the rates of disproportionation for HOO·, its relative rate of oxidation of 1,4-CHD in the three solvents is in the same order, PhCl>MeCN>Me$_2$SO. For Me$_2$SO the value of $k_{ox}$ is slightly greater than $k_d^{1/2}$, but with MeCN and PhCl the values of $k_{ox}$ are an order of magnitude smaller than their $k_d^{1/2}$ values. Hence, the propensity of HOO· to initiate the auto-oxidation of allylic groups is greater in Me$_2$SO and occurs at much lower concentrations of substrate. The rate constants $(k_{ox})$ for the allylic-hydrogen oxidation of linoleic acid and arachidonic acid in protic media also are slightly greater than $(k_d^{1/2})_{H_2O}$ (Tables 5-3 and 5-4). The large rate constants $(k_{ox})$ for the reaction of HOO· with antioxidants ($\alpha$-tocopherol, cysteine, and ascorbic acid) is consistent with their protective role in biology.

Table 5-4 Rate Constants for the Second-Order Oxidation by HOO· of Allylic and Other X–H Functions of Organic Substrates $(k_{ox})$

| Substrate | Solvent | $k_{ox}$ ($M^{-1}$ $s^{-1}$) | Ref. |
|---|---|---|---|
| 1,4-CHD | Me$_2$SO | $1.6 \times 10^2$ | 21 |
| | MeCN | $3.5 \times 10^2$ | 22 |
| | PhCl | $1.4 \times 10^3$ | 22 |
| linoleic acid | 85:15 EtOH/H$_2$O | $1.2 \times 10^3$ | 23 |
| arachidonic acid | 85:15 EtOH/H$_2$O | $3.0 \times 10^3$ | 23 |
| $\alpha$-tocopherol | 85:15 EtOH/H$_2$O | $2.0 \times 10^5$ | 24 |
| cysteine | H$_2$O | $1.8 \times 10^4$ | 25 |
| ascorbic acid | H$_2$O | $1.6 \times 10^4$ | 26 |

A likely biological function for the superoxide dismutase proteins (SOD) is to remove $O_2^-$·, and thereby preclude formation of HOO· [Eq. (5-18)] and prevent initiation of lipid peroxidation and autoxidation (Scheme 5-2). An SOD model complex $[Mn^{II}(PA)_2]$ effectively blocks formation of HOO· from $O_2^{-}$·, and prevents reaction with 1,4-CHD.[27]

*Scheme 5-2   Formation and reactivity of HOO·*

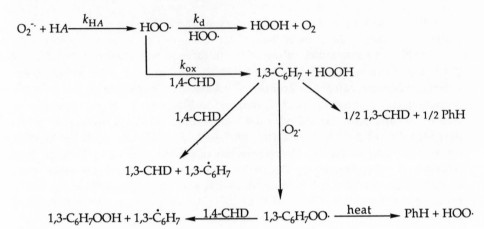

The reactivity of ROO· radicals parallels that of HOO·, and is essentially equivalent because its bond energy (ROO–H; $-\Delta G_{BF}$, 83 kcal) is the same as that for the HOO–H bond ($-\Delta G_{BF}$, 82 kcal). However, the "head-to-tail" mechanism for the disproportionation of HOO· [via H-atom transfer, Eq. (5-21)] is not possible. Instead ROO· radicals dimerize via radical–radical coupling to form a dialkyl tetraoxide.[28]

$$2\ ROO· \xrightarrow{\ k_d\ } [ROOOOR] \longrightarrow ROOR + {}^1O_2 \qquad (5\text{-}23)$$

The [ROOOOR] intermediate decomposes to dialkyl peroxide and singlet dioxygen, but also reacts more vigorously than $^1O_2$ with "$^1O_2$ substrates" (1,4-CHD, 1,3-diphenylisobenzofuran, rubrene; Chapter 6).

## References

1. Sawyer, D. T. *J. Phys. Chem.* **1990**, *93*, 7977.
2. Lide, D. R., (ed.). *Handbook of Chemistry and Physics*, 71st ed. Boca Raton, FL: CRC, 1990, pp. 9-86–98.
3. Atkinson, R. *Chem. Rev.* **1986**, *86*, 69.
4. Matheson, M. S.; Dorfman, L. M. *Pulse Radiolysis*. Cambridge: M.I.T. Press, 1969.
5. (a) Dorfman, L. F.; Adams, G. E. *Natl. Stand. Ref. Data Serv. (U. S., Natl. Bur. Stand.)* **1973**, NSRDS-NBS46, 20 (SD catalog No. C13.48:46); (b) Burton, G. V.; Greenstock, C. L.; Helman, W. D.; Ross, A. B. *J. Phys. Chem. Ref. Data* **1988**, *17*, 513.
6. Walling, C. *Acc. Chem. Res.* **1975**, *8*, 125.

7. Hardwick, T. J. *Can. J. Chem.* **1957**, *35*, 428.
8. Rush, J. D.; Koppenol, W. H. *J. Am. Chem. Soc.* **1988**, *110*, 4957.
9. Sheu, C.; Sobkowiak, A.; Zhang, L.; Ozbalik, N.; Barton, D. H. R.; Sawyer, D. T. *J. Am. Chem. Soc.* **1989**, *111*, 8030.
10. Roberts, J. L., Jr.; Morrison, M. M; Sawyer, D. T. *J. Am. Chem. Soc.* **1978**, *100*, 329.
11. (a) Minisci, F.; Citterio, A.; Giordano, C. *Acc. Chem. Res.* **1983**, *16*, 27; (b) Minisci, F.; Citterio, A.; Vismara, E.; Giordano, C. *Tetrahedron* **1985**, *41*, 4157.
12. Sheu, C.; Richert, S. A.; Cofré, P.; Ross, B., Jr.; Sobkowiak, A.; Sawyer, D. T.; Kanofsky, J. R. *J. Am. Chem. Soc.* **1990**, *112*, 1936.
13. Trotman-Dickenson, A. F. *Adv. Free Radical Chem.* **1965**, *1*, 1.
14. Bielski, B. H. J.; Cabelli, D. E.; Arudi, R. L.; Ross, A. B. *J. Phys. Chem. Ref. Data.* **1985**, *14*, 1041.
15. Sawyer, D. T.; Valentine, J. S. *Acc. Chem. Res.* **1981**, *14*, 393.
16. Nadezhdin, A.; Dunford, H. B. *J. Phys. Chem.* **1979**, *83*, 1957.
17. Fridovich, I. In *Advances in Inorganic Biochemistry* (Eichhorn, G. L.; Marzilli, D. L., eds.). New York: Elsevier–North Holland, 1979, pp. 67–90.
18. Chin, D.-H.; Chiericato, G., Jr.; Nanni, E. J., Jr.; Sawyer, D. T. *J. Am. Chem. Soc.* **1982**, *104*, 1296.
19. Cofré, P.; Sawyer, D. T. *Inorg. Chem.* **1986**, *25*, 2089.
20. Sugimoto, H.; Matsumoto, M.; Kanofsky, J. R.; Sawyer, D. T. *J. Am. Chem. Soc.* **1988**, *22*, 1182.
21. Sawyer, D. T.; McDowell, M. S.; Yamaguchi, K. S. *Chem. Res. Toxicol.* **1988**, *1*, 97.
22. Howard, J. A.; Ingold, K. V. *Can. J. Chem.* **1967**, *45*, 785.
23. Bielski, B. H. J.; Arudi, R. L.; Sutherland, M. W. *J. Biol. Chem.* **1983**, *258*, 4759.
24. Arudi, R. L.; Sutherland, M. W.; Bielski, B. H. J. In *Oxy Radicals and Their Scanvenger Systems* (Cohen, G.; Greenwald, R. A., eds.), Vol. 1. New York: Elsevier Biochemical, 1983, pp. 26–31.
25. Al-Thannon, A. A.; Barton, J. P.; Packer, J. E.; Sims, R. J.; Trumbore, C. N.; Winchester, R. V. *Int. J. Radiat. Phys. Chem.* **1974**, *6*, 233.
26. Cabelli, D. E.; Bielski, B. H. J. *J. Phys. Chem.* **1983**, *87*, 1809.
27. Yamaguchi, K. S.; Spencer, L.; Sawyer, D. T. *FEBS Lett.* **1986**, *197*, 249.
28. Kanofsky, J. R.; Matsumoto, M.; Sugimoto, H.; Sawyer, D. T. *J. Am. Chem. Soc.* **1988**, *110*, 5193.

# 6

# REACTIVITY OF DIOXYGEN AND ITS ACTIVATION FOR SELECTIVE DIOXYGENATION, MONO-OXYGENATION, DEHYDROGENATION, AND AUTO-OXIDATION OF ORGANIC SUBSTRATES AND METALS (CORROSION)

$O_2$: *The basis of aerobic life and biological combustion*

## Introduction

Ground-state dioxygen has two unpaired electrons ($\cdot O_2 \cdot$), which makes it a biradical with a triplet electronic state (see Table 3-1). Its radical character is limited because the H–OO· bond is weak [$-\Delta G_{BF}$, 51 kcal (Chapter 3)], and the triplet state precludes direct reaction with singlet-state substrate molecules with saturated $\sigma$ bonding.

Perhaps the most important (but nonproductive) reaction chemistry for $^3O_2$ is its reversible binding by the metalloproteins: hemoglobin, myoglobin, hemerythrin, and hemocyanin. Nature developed such systems to obviate the limited solubility of $O_2$ in water (~1 m$M$ at 1 atm $O_2$), which restricts the energy flux from oxidative metabolism in aerobic organisms. In the case of myoglobin (a reduced heme protein with an axial histidine base; :B), the reversible binding of $\cdot O_2 \cdot$ causes the paramagnetic iron(II) center ($S = 4/2$) of the heme to become diamagnetic via the formation of two covalent Fe–($O_2$) bonds (the estimated charge transfer for the iron to the bound $O_2$ is about 0.1 electron),[1]

$$B\!:\!\overset{|}{\underset{|}{Fe}}^{II} + O_2 \rightleftharpoons B\!:\!\overset{|}{\underset{|}{Fe}}^{IV}\!\!\overset{O}{\underset{\cdots\cdots}{\diagdown}}O \qquad K_{O_2} = \frac{1}{(P_{1/2})_{O_2}} \approx 10^3 \, \text{atm}^{-1} \qquad (6\text{-}1)$$

$(d^6sp, S = 4/2)$  $(d^6sp, S=0)$

where $(P_{1/2})_{O_2}$ represents the partial pressure of $O_2$ when one-half of the heme centers have an $O_2$ adduct. Carbon monoxide is more strongly bound, with a $K_{CO}$ value of about $10^6$ atm$^{-1}$ via an analogous nucleophilic interaction by the reduced imidiazole-liganded iron–porphyrin center ($B\!:\!\overset{|}{\underset{|}{Fe}}^{IV}$=C=O, $d^6sp$, $S=0$).

Although the function of the $O_2$-binding proteins is the enhancement of the $O_2$ concentration for transport and storage, there is an inevitable degradation of the oxy–heme system, primarily via interaction with nonoxygenated reduced hemes to give ($B\!:\!\overset{|}{\underset{|}{Fe}}^{III}O\overset{|}{\underset{|}{Fe}}^{III}\!:\!B$). The kinetics and mechanism for this irreversible process have been characterized through the use of hindered ("picket-fence") iron–porphyrin models for myoglobin.[2]

$$B:Fe^{IV}(O_2) + B:Fe^{II} \longrightarrow B:Fe^{III}OOFe^{III}:B$$

oxy-myoglobin

$$\underset{\text{2 B:Fe}}{\overset{}{\big\downarrow}} \quad 2\ B:Fe^{III}OFe^{III}:B \qquad (6\text{-}2)$$

met-myoglobin

The "picket fence" hinders the face-to-face approach of the two hemes, but less effectively than the polypeptide chains of the heme protein. The auto-oxidation pathway for the heme in myoglobin [Eqs. (6-1) and (6-2)] is sufficiently slow to give the protein a useful life of several weeks, but is analogous to that for all reduced-iron systems.

Biological systems overcome the inherent unreactive character of $^3O_2$ by means of metalloproteins (enzymes) that activate dioxygen for selective reaction with organic substrates. For example, the cytochrome $P$-450 proteins (thiolated protoporphyrin IX catalytic centers) facilitate the epoxidation of olefins, the demethylization of N-methyl amines (via formation of formaldehyde), the oxidative cleavage of $\alpha$-diols to aldehydes and ketones, and the monooxygenation of aliphatic and aromatic hydrocarbons (RH).[3]

$$O_2 + (2\ H^+ + 2\ e^-) + RH \xrightarrow{\text{cyt } P\text{-450}} ROH + H_2O \qquad (6\text{-}3)$$
cytochrome $P$-450 reductase

The methane mono-oxygenase proteins (MMO, binuclear nonheme iron centers) catalyze similar oxygenation of saturated hydrocarbons[4,5]

$$O_2 + (2\ H^+ + 2\ e^-) + CH_4 \xrightarrow{\text{MMO}} CH_3OH + H_2O \qquad (6\text{-}4)$$
reductase

Both of these systems activate $O_2$ via a two-equivalent reduction (without the protein catalysts this would give HOOH). In contrast, pyrocatechol dioxygenase (PDO) (a nonheme iron protein)[6] activates $^3O_2$ without a reductive co-factor

$$\text{(catechol)} + O_2 \xrightarrow{\text{PDO}} HOC(O)CH=CH-CH=CHC(O)OH \qquad (6\text{-}5)$$
muconic acid

The most direct means to activate $^3O_2$ is by reduction with electrons or hydrogen atoms to give $O_2^-$, HOO·, HOOH, and HOO$^-$. Chapter 7 discusses the reaction chemistry for $O_2^-$, and Chapter 5 is devoted to the reactivity of oxygen radicals (including HOO·). The chemistry of HOOH is covered in Chapter 4, and that of HOO$^-$ in Chapter 8.

Ground-state dioxygen ($^3O_2$) also can be activated by photochemical energy transfer to yield singlet dioxygen $^1O_2$.[7]

$$^3O_2 \xrightarrow[\text{Rose Bengal}]{h\upsilon} {}^1O_2 \qquad\qquad (6\text{-}6)$$

The chemical reactivity of $^1O_2$ is discussed at the end of this chapter.

### Radical–radical coupling and auto-oxidation

Although $^3O_2$ is unreactive with singlet substrates and is unable to abstract hydrogen atoms from hydrocarbon substrates, it is a biradical and as such can undergo coupling with other radicals (carbon and nonmetals as well as transition metals). Such radical–radical coupling reactions usually have zero activation energy. Thus, an HO· radical can initiate the reaction of organic substrates with $^3O_2$.

*Initiation*:

$$\text{PhCH(O)} + \text{HO·} \longrightarrow \text{Ph}\underset{.}{\text{C}}\text{(O)} + \text{HOH} \qquad\qquad (6\text{-}7a)$$

*Propagation*:

$$\text{Ph}\underset{.}{\text{C}}\text{(O)} + \cdot O_2\cdot \longrightarrow \text{PhC(O)OO·} \qquad\qquad (6\text{-}7b)$$

$$\text{PhC(O)OO·} + \text{PhCH(O)} \longrightarrow \text{Ph}\underset{.}{\text{C}}\text{(O)} + \text{PhC(O)OOH} \qquad\qquad (6\text{-}7c)$$

*Termination*:

$$\text{PhC(O)OOH} + \text{PhCH(O)} \longrightarrow 2\,\text{PhC(O)OH} \qquad\qquad (6\text{-}7d)$$

Propagation in such autoxidation cycles by $^3O_2$ requires that the $ROO·$ intermediate be a sufficiently strong radical to break a C–H bond of the substrate. Allylic, aldehydic, and benzylic C–H bonds are examples that meet this limitation.

Reduced transition metals with unpaired electrons also undergo auto-oxidation via radical–radical coupling. For example, reduced iron porphyrins are rapidly converted to binuclear μ-oxo dimers.[8]

$$\text{Fe}^{II}(\text{TPP}) + O_2 \rightleftharpoons (\text{TPP})\text{Fe}^{IV}(O_2) \qquad\qquad (6\text{-}8)$$
$$(d^6sp,\ S = 4/2)$$

$$\downarrow \text{Fe}^{II}(\text{TPP})$$

$$(\text{TPP})\text{Fe}^{III}\text{OOFe}^{III}(\text{TPP})$$

$$\downarrow 2\,\text{Fe}^{II}(\text{TPP})$$

$$2\,(\text{TPP})\text{Fe}^{III}\text{OFe}^{III}(\text{TPP})$$
$$(d^5sp^2,\ S=5/2)$$

Similar heterogeneous radical–radical coupling occurs at the solution–surface interface of metallic iron.

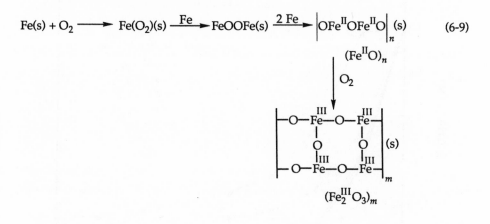

$$Fe(s) + O_2 \longrightarrow Fe(O_2)(s) \xrightarrow{\ Fe\ } FeOOFe(s) \xrightarrow{\ 2\,Fe\ } \left| O Fe^{II} O Fe^{II} O \right|_n (s) \qquad (6\text{-}9)$$

$$(Fe^{II}O)_n$$

$$\downarrow O_2$$

$$(Fe_2^{III}O_3)_m$$

## Metal-induced activation of $^3O_2$ for the initiation of auto-oxidation

Reduced transition-metal complexes are traditionally implicated as the initiators for the auto-oxidation of fats, lipids, and foodstuffs.[9] However, whether this involves direct activation of $O_2$ or of reduced dioxygen ($O_2^{-\cdot}$, HOO·, and HOOH) is unclear. Reduced metal plus HOOH yields HO· via Fenton chemistry and probably is the pathway for initiation of auto-oxidation in many systems (see Chapter 4). Although there has been an expectation that one or more of the intermediates from the auto-oxidation of reduced transition metals [Eq. (6-8)] can act as the initiator for the auto-oxidation of organic substrates, direct experimental evidence has not been presented.

The addition of trace levels ($>10^{-6}\,M$) of bis(bipyridine)cobalt(II) to $O_2$-saturated solutions of aldehydes in acetonitrile initiates their rapid auto-oxidation to carboxylic acids.[10] Figure 6-1 illustrates the $Co^{II}(bpy)_2^{2+}$-induced auto-oxidation of hexanal [$CH_3(CH_2)_4CH(O)$] for $O_2$-saturated (8.1 m$M$) and air-saturated (1.6 m$M$) acetonitrile. The apparent reaction dynamics for the catalyzed auto-oxidation of PhCH(O) and of $CH_3(CH_2)_4CH(O)$ during the first hour of their auto-oxidation is summarized in Table 6-1. The initial reaction rates appear to be first order in catalyst concentration, first order in substrate concentration, and first order in $O_2$ concentration (Fig. 6-1). However, within one hour the auto-oxidation process is almost independent of catalyst concentration. Although the $Fe^{II}(bpy)_2^{2+}$ and $Mn^{II}(bpy)_2^{2+}$ complexes also induce the auto-oxidation of aldehydes, they are much less effective initiators, and the propagation dynamics are much slower.

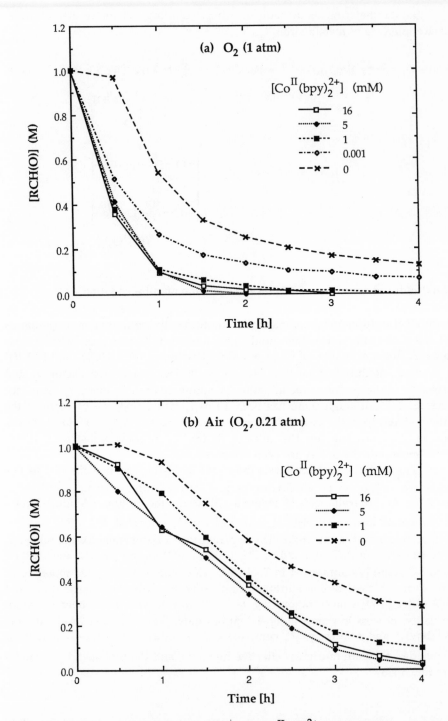

Figure 6-1 The decay of $CH_3(CH_2)_4CH(O)$ via its $Co^{II}(bpy)_2^{2+}$-induced auto-oxidation in MeCN in the presence of (a) dioxygen (1 atm) and (b) air. For 1 mM $Co^{II}(bpy)_2^{2+}$ and $O_2$ (1 atm) the half-life of $RCH(O)$ is 20 min; with air, $t_{1/2}$ is 140 min.

Table 6-1 $Co^{II}(bpy)_2^{2+}$-Induced Auto-oxidation of Aldehydes in Acetonitrile Solutions

| Substrate (1 M) | $Co^{II}(bpy)_2^{2+}$ (mM) | $O_2$ (mM) | $k_{obs}$ $(M^{-1} s^{-1})^a$ | Products (mM after 1 h) |
|---|---|---|---|---|
| PhCH(O) | 0.0 | 8.1 | $0.5 \times 10^{-3}$ | PhC(O)OH (15) |
| PhCH(O) | 0.001 | 8.1 | $1.7 \times 10^{-3}$ | PhC(O)OH (180) |
| PhCH(O) | 1.0 | 8.1 | $5.7 \times 10^{-3}$ | PhC(O)OH (170) |
| PhCH(O) | 5.0 | 8.1 | $11.4 \times 10^{-3}$ | PhC(O)OH (315) |
| PhCH(O) | 16.0 | 8.1 | $9.7 \times 10^{-3}$ | PhC(O)OH (255) |
| PhCH(O) | 0.0 | 1.6 | $12.5 \times 10^{-3}$ | PhC(O)OH (70) |
| PhCH(O) | 1.0 | 1.6 | $26.0 \times 10^{-3}$ | PhC(O)OH (150) |
| PhCH(O) | 5.0 | 1.6 | $17.5 \times 10^{-3}$ | PhC(O)OH (80) |
| PhCH(O) | 16.0 | 1.6 | $26.0 \times 10^{-3}$ | PhC(O)OH (150) |
| $CH_3(CH_2)_4CH(O)$ | 0.0 | 8.1 | 0 | $CH_3(CH_2)_4C(O)OH$ (300) |
| $CH_3(CH_2)_4CH(O)$ | 0.001 | 8.1 | $15.6 \times 10^{-3}$ | $CH_3(CH_2)_4C(O)OH$ (510) |
| $CH_3(CH_2)_4CH(O)$ | 1.0 | 8.1 | $18.7 \times 10^{-3}$ | $CH_3(CH_2)_4C(O)OH$ (540) |
| $CH_3(CH_2)_4CH(O)$ | 5.0 | 8.1 | $15.9 \times 10^{-3}$ | $CH_3(CH_2)_4C(O)OH$ (450) |
| $CH_3(CH_2)_4CH(O)$ | 16.0 | 8.1 | $15.6 \times 10^{-3}$ | $CH_3(CH_2)_4C(O)OH$ (450) |
| $CH_3(CH_2)_4CH(O)$ | 0.0 | 1.6 | 0 | $CH_3(CH_2)_4C(O)OH$ (25) |
| $CH_3(CH_2)_4CH(O)$ | 1.0 | 1.6 | $20.1 \times 10^{-3}$ | $CH_3(CH_2)_4C(O)OH$ (105) |
| $CH_3(CH_2)_4CH(O)$ | 5.0 | 1.6 | $26.0 \times 10^{-3}$ | $CH_3(CH_2)_4C(O)OH$ (150) |
| $CH_3(CH_2)_4CH(O)$ | 16.0 | 1.6 | $22.0 \times 10^{-3}$ | $CH_3(CH_2)_4C(O)OH$ (140) |

$^a$ $k_{obs} = \{-d[RC(O)OH]/dt\}/[RCH(O)][O_2]_{t=0}$

## Metal-induced activation of dioxygen for oxygenase and dehydrogenase chemistry

*Direct ketonization of methylenic carbons, and the dioxygenation of aryl olefins, acetylenes, and catechols*

Chapter 4 describes the activation of HOOH by bis(picolinato)iron(II) [$Fe^{II}(PA)_2$] to form $(PA)_2Fe^{III}OOFe^{III}(PA)_2$, which selectively reacts with (1) methylenic carbons ( $\diagdown$CH$_2$) to form ketones ( $\diagdown$C=O), (2) arylolefins to form aldehydes [$c$-PhCH=CHPh $\longrightarrow$ 2 PhCH(O)], (3) acetylenes to form $\alpha$-diones, and (4) catechol to form muconic acid. Although $Fe^{II}(PA)_2$ is almost inert to auto-oxidation by dioxygen, the bis(2,6-carboxyl, carboxylato-pyridine)iron(II) complex [$Fe^{II}(DPAH)_2$] is rapidly auto-oxidized by $O_2$ [1 atm, 3.4 m$M$ in py/HOAc (2:1)] to an intermediate that has analogous substrate reactivity to the $(PA)_2Fe^{III}OOFe^{III}(PA)_2$ intermediate from the $Fe^{II}(PA)_2$/HOOH system.[11]

The products and reaction efficiencies for various concentrations of catalysts [$Fe^{II}(DPAH)_2$] and substrates are summarized in Table 6-2. The products are identical to those that result from the reactive intermediates that are formed from the combination of $(DPA)Fe^{III}OFe^{III}(DPA)$, $(PA)_2Fe^{III}OFe^{III}(PA)_2$, $Fe^{II}(PA)_2$, or $Fe^{II}(DPA)$ with excess HOOH, for example, [$(DPA)Fe^{III}OOFe^{III}(DPA)$] (see Chapter 4).[12] The dioxygenation of the substrates in Table 6-2 must involve a similar reactive intermediate. With cyclohexane ($c$-$C_6H_{12}$) about one-fourth of the $O_2$ that is incorporated into this reactive intermediate reacts to give cyclohexanone as the only detectable product; the remainder oxidizes excess $Fe^{II}(DPAH)_2$ to give $(DPAH)_2Fe^{III}OFe^{III}(DPAH)_2$, which is catalytically inert. This chemistry is outlined in Scheme 6-1.

*Scheme 6-1 Activation of $O_2$ by $Fe^{II}(DPAH)_2$ in 1.8:1 py/HOAc*

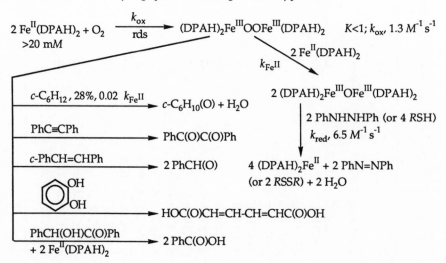

Table 6-2    Ketonization of Methylenic Carbons and Dioxygenation of Aryl Olefins, Acetylenes, and Catechols via the $Fe^{II}(DPAH)_2$-Induced Activation of Dioxygen in 1.8:1 py/HOAc[a]

| Substrate | Product (mM)[b] | Reaction efficiency [% (±3)][c] |
|---|---|---|
| $c$-$C_6H_{12}$ (1 $M$) | $c$-$C_6H_{12}$(O) (4.5) | 28 |
| $PhCH_2CH_3$ (1 $M$) | $PhC(O)CH_3$ (3.5) | 22 |
| [+128 mM PhNHNHPh] | $PhC(O)CH_3$ [18.9] | |
| 2-Me-butane (1 $M$) | $Me_2CHC(O)Me$ (1.0) | 6 |
| [+128 mM PhNHNHPh] | $Me_2CHC(O)Me$ [9.1] | |
| Cyclohexene (1 $M$) | 2-cyclohexene-1-one (1.2) | 7 |
| $PhC{\equiv}CPh$ (0.6 $M$) | $PhC(O)C(O)Ph$ (2.2) | 14 |
| $c$-PhCH=CHPh (1 $M$) | PhCH(O) (3.1) | 10 |
| 1,2-$Ph(OH)_2$ (1 $M$) | HOC(O)CH=CH-CH=CHC(O)OH | 13 |
| | (and its anhydride) (2.0) | |
| $PhCH(OH)C(OH)Ph$ (0.3 $M$) | $PhC(O)OH$ (5.2) | 16 |
| PhNHNHPh (100 mM) | PhN=NPh (100) | 667[d] |
| $PhCH_2SH$ (128 mM) | $PhCH_2SSCH_2Ph$ (64) | 800[d] |
| $H_2S$ (128 mM) | $S_8$ (16.0) | 800[d] |

[a] $Fe^{II}(DPAH)_2$ (32 mM); $O_2$ (1 atm, 3.4 mM). Substrate and $Fe^{II}(DPAH)_2$ [$Fe(MeCN)_4(ClO_4)_2$ added to two equivalents of $(Me_4N)_2DPA$] combined in 3.5 mL of pyridine/HOAc solvent (1.8:1 mol ratio), followed by the addition of $O_2$ (1 atm, 3.4 mM) in a reaction cell with 6 mL of head space. Reaction time and temperature: 4 h at 22±2 °C [for 3 mM $Fe^{II}(DPAH)_2$, the reaction time was 12 h].

[b] The product solutions were analyzed by capillary gas chromatography and GC-MS (either direct injection of the product solution, or by quenching with water and extracting with diethyl either).

[c] 100% represents one substrate ketonization or dioxygenation per $(DPAH)_2FeOOFe(DPAH)_2$ reactive intermediate.

[d] 100% represents one substrate oxidation per $(DPAH)_2FeOFe(DPAH)_2$ reactive intermediate.

*Catalyzed dehydrogenation of hydrazines, thiols, and hydrogen sulfide*

In the absence of substrate the active catalyst $[Fe^{II}(DPAH)_2]$ is rapidly auto-oxidized to $(DPAH)_2Fe^{III}OFe^{III}(DPAH)_2$ [4 $Fe^{II}(DPAH)_2$ per $O_2$; the apparent second-order rate constant, $k_{ox}$, has a value of 1.3±0.5 $M^{-1}$ $s^{-1}$ ($k_{obs}/4$)]. The oxidized catalyst $[(DPAH)_2Fe^{III}OFe^{III}(DPAH)_2]$ is rapidly reduced to $Fe^{II}(DPAH)_2$ by PhNHNHPh, $H_2NNH_2$, $PhCH_2SH$, and $H_2S$ ($k_{red}$, 6.5±0.5, 0.6±0.3, 0.5±0.3, and 2.8±0.5 $M^{-1}$ $s^{-1}$, respectively) to give PhN=NPh, $N_2$, $PhCH_2SSCH_2Ph$, and elemental sulfur ($S_8$), respectively (Table 6-2).[11] The PhNHNHPh reductant is an effective reaction mimic for the reduced flavin cofactors in xanthine oxidase and cytochrome $P$-450 reductase.[13]

The catalytic cycle for the auto-oxidation of mecaptans and $H_2S$ (Scheme 6-1) is general for several transition-metal complexes. Industrial processes for the removal of $RSH$ and $H_2S$ from gas streams make use of $Fe^{II}(EDTA)$ and $VO-$ (catechol)$_2^{2-}$ catalysts in aqueous solutions that are saturated with air (21 mol % $O_2$).

The results of Table 6-2 and the close parallels of the product profiles to those for the $(DPA)Fe^{III}OFe^{III}(DPA)/HOOH/(py/HOAc)$ system (Chapter 4)[12] prompt the conclusion that the combination of $Fe^{II}(DPA)_2$ and $O_2$ results in the initial formation of the reactive intermediate $[(DPAH)_2Fe^{III}OOFe^{III}(DPAH)_2, 1]$ via a rate-limiting step (Scheme 6-1), and are the basis for the proposed reaction pathways. Because $K$ is less than unity, the yield of cyclohexanone increases linearly with $Fe^{II}(DPAH)_2$ concentration. The apparent rate for the ketonization reaction is proportional to substrate concentration, $Fe^{II}(DPAH)_2$ concentration, and $O_2$ partial pressure, and increases with temperature (about five times faster at 25°C than at 0°C). Because the fraction of 1 that reacts with $c$-$C_6H_{12}$ remains constant (~28%), the oxidation of excess $Fe^{II}(DPAH)_2$ by 1 must be a parallel process. Given that the ratio of concentrations $[c$-$C_6H_{12}]/[Fe^{II}(DPAH)_2]$ is about 30:1 and the ratio of reactivities is 1:2.6, the apparent relative rate constant for reaction of $c$-$C_6H_{12}$ and $Fe^{II}(DPAH)_2$ with 1 is about 0.02 $k_{Fe^{II}}$ (assuming a stoichiometric factor of 2 for the latter).

Addition of $PhNHNHPh$, $H_2NNH_2$, $PhCH_2SH$, or $H_2S$ to the reaction system $[O_2/Fe(DPAH)_2/substrate$ in 1.8:1 py/HOAc] reduces the oxidized catalyst $[(DPAH)_2FeOFe(DPAH)_2]$ and thereby recycles it for activation of $O_2$ to the reactive intermediate (Table 6-2). When 3 m$M$ $Fe^{II}(DPAH)_2$ is used in combination with 100 m$M$ $PhNHNHPh$, the rate for the ketonization of $c$-$C_6H_{12}$ is reduced by an order of magnitude, but each cycle remains about 21% efficient with 67 turnovers within 12 h.

The dioxygenation of unsaturated $\alpha$-diols (catechol and benzoin, Table 6-2) by the $O_2/Fe^{II}(DPAH)_2$ system parallels that of the catechol dioxygenase enzymes, which are nonheme iron proteins.[6,14] Hence, the reactive intermediate (1, Scheme 6-1) of the $Fe^{II}(DPAH)_2/O_2$ reaction may be a useful model and mimic for the activated complex of dioxygenase enzymes.[6]

This system affords the means to the selective auto-oxidation (oxygenation) of hydrocarbon substrates (e.g., $c$-$C_6H_{12}$) via the coprocessing of $H_2S$ (or $RSH$) -contaminated hydrocarbon streams. Thus, the combination of $c$-$C_6H_{12}$ and $H_2S$ with $Fe^{II}(DPAH)_2$ and $O_2$ in 1.8:1 py/HOAc yields $c$-$C_6H_{10}(O)$ and $S_8$.

*Demethylization of N-methyl anilines, and the dehydrogenation of $PhCH_2NH_2$ and $PhCH_2OH(+ HO^-)$*

The addition of $Co^{II}(bpy)_2^{2+}$ to dioxygen-saturated acetonitrile solutions of N-methyl anilines catalyzes their demethylization via a mono-oxygenase pathway to give formaldehyde and demethylated aniline.[15] This is analogous to the chemistry facilitated by cytochrome $P$-450 proteins with N-methyl anilines.[3] The $Co^{II}(bpy)_2^{2+}/O_2/MeCN$ system also dehydrogenates benzylamines and basic benzyl alcohol.

$$2\ PhCH_2NH_2 + O_2 \xrightarrow{\ Co^{II}(bpy)_2^{2+}\ } 2\ PhCH=NH + 2\ H_2O \qquad (6\text{-}10)$$

$$\xrightarrow[\phantom{xx}]{H_2O} 2\ PhCH(O) + 2\ NH_3$$

$$2\ PhCH_2OH + O_2 \xrightarrow{\ (HO)Co^{II}(bpy)_2^{+}\ } 2\ PhCH(O) + 2\ H_2O \qquad (6\text{-}11)$$

Table 6-3 summarizes the product yields for these substrates in $O_2$-saturated acetonitrile. The rates of transformation are first order in $Co^{II}(bpy)_2^{2+}$ con-

Table 6-3 $Co^{II}(bpy)_2^{2+}$-Induced Activation of $O_2$ (1 atm, 8.1 m$M$) in Acetonitrile for (a) the Demethylization of N-Methyl Anilines and (b) the Mono-oxygenation of $PhCH_2NH_2$ and $PhCH_2OH(+ HO^-)$.

a. N-Methyl Anilines; $Co^{II}(bpy)_2^{2+}$ (16 m$M$); 24-h reaction time

| Substrate (1 $M$) | Products (1 m$M$) |
|---|---|
| $PhNH_2$ | N R |
| PhNHMe | $PhNH_2$ (16), $H_2C(O)$ |
| $PhN(Me)_2$ | PhNHMe (64), dimer (34), $H_2C(O)$ |
| $PhCH_2N(Me)_2$ | N R |

b. $PhCH_2NH_2$ and $PhCH_2OH(+HO^-)$; $(HO)Co^{II}(bpy)_2^{2+}$ (16 m$M$); 24-h reaction time

| Substrate (1 $M$) | Products (m$M$) |
|---|---|
| $PhCH_2NH_2$[a] | $PhCH_2N=CHPh$ (8) |
| $PhCH_2NHMe$[b] | $PhCH_2N=CHPh$ (6) |
| $PhCH_2NHPh$[c] | $PhN=CHPh$ (7) |
| $PhCH_2N(Me)_2$ | N R |
| $(PhCH_2)_2NH$ | N R |
| $PhCH_2OH$ | $PhCH(O)$ (91) |

[a] Combination of 16 m$M$ $Co^{II}(bpy)_2^{2+}$, 200 m$M$ HOOH, and 1 $M$ $PhCH_2NH_2$ for 12 h yields 70 m$M$ $PhCH_2N=CHPh$.

[b] Combination of 16 m$M$ $Co^{II}(bpy)_2^{2+}$, 200 m$M$ HOOH, and 1 $M$ $PhCH_2NHCH_3$ for 12 h yields 11 m$M$ $PhCH_2N=CHPh$.

[c] Combination of 16 m$M$ $Co^{II}(bpy)_2^{2+}$, 200 m$M$ HOOH, and 1 $M$ $PhCH_2NHPh$ for 12 h yields 16 m$M$ $PhCH(O)$, 16 m$M$ $PhNH_2$, and 3 m$M$ $PhN=CHPh$.

centration, first order in $O_2$ concentration, and first order in substrate concentration.

The combination of $Co^{II}(bpy)_2^{2+}$-$O_2$ with these substrates yields product profiles that are similar, which indicates that the reaction path probably involves the same reactive intermediate (species **2**, Scheme 6-2). Scheme 6-2 outlines catalytic cycles for the demethylation of N-methyl anilines, and the dehydrogenation of benzylamine and benzyl alcohols. In the case of the latter catalyst $[Co^{II}(bpy)_2^{2+}]$ is inactive and must be neutralized with one equivalent of $HO^-$

$$\cdot Co^{II}(bpy)_2^{2+} + HO^- \longrightarrow HOC\underset{\cdot}{o}^{II}(bpy)_2^+ \qquad (6\text{-}12)$$

*Scheme 6-2   Demethylation of N-Methyl Anilines, and Dehydrogenation of Benzylamine and Benzyl Alcohol*

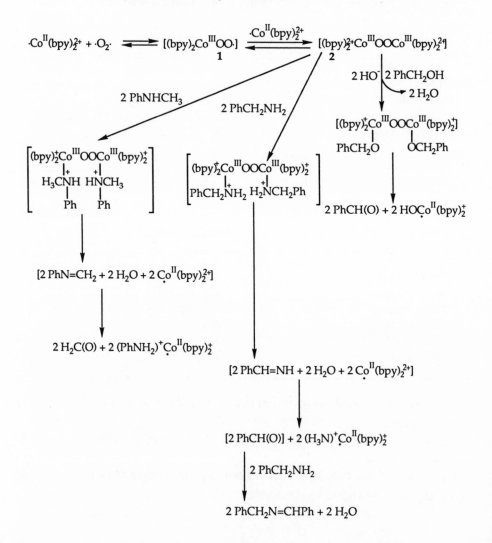

Either $HOC(O)O^-$ or $Et_3N\colon$ can be used in place of $HO^-$, but the reaction rates are slower. In the absence of substrate, the neutralized catalyst is autoxidized via residual water in the solvent to generate HOOH.

$$2\,HOCo^{II}(bpy)_2^+ + O_2 + 2\,H_2O \longrightarrow 2\,(HO)_2Co^{III}(bpy)_2^+ + HOOH \qquad (6\text{-}13)$$

$$\Big\downarrow 2\,H_2O$$

$$2\,Co^{III}(OH)_3\,(s) + 2\,H(bpy)_2^+$$

**Metal-induced reductive activation of $O_2$ for mono-oxygenation**

*Reaction mimic for the cytochrome P-450 mono-oxygenase/reductase system*
*[$Fe^{II}(MeCN)_4^{2+}$ /$O_2$/PhNHNHPh/substrate]*

Chapter 4 discusses the ability of $Fe^{II}(MeCN)_4^{2+}$ to activate HOOH for mono-oxygenase and dehydrogenase chemistry, for example,

$$Fe^{II}(MeCN)_4^{2+} + HOOH \longrightarrow (MeCN)_4^{2+}Fe^{II}\!\left(\begin{matrix}H\\O\\O\\H\end{matrix}\right) \xrightarrow{Ph_2S(O)} \qquad (6\text{-}14)$$

$$Ph_2S(O)_2 + (H_2O)^+Fe^{II}(MeCN)_4^+$$

A separate study[13] has shown that PhNHNHPh is an effective reaction mimic for the flavin cofactors in xanthine oxidases and cytochrome P-450 reductases, and in combination with $O_2$ yields HOOH.

$$PhNHNHPh + O_2 \xrightarrow{\text{base}} PhN{=}NPh + HOOH \qquad (6\text{-}15)$$

Table 6-4 summarizes the product yields from the combination of substrate [$Ph_3P$, $Ph_2S(O)$, $PhCH_2OH$, or 1,4-cyclohexadiene] with $O_2$, PhNHNHPh, and $Fe^{II}(MeCN)_4^{2+}$ in dimethylformamide.[15] The system acts as an effective mono-oxygenase/dehydrogenase, and as such mimics the reaction cycle of the cytochrome P-450 mono-oxygenase/reductase system.

$$Ph_2S(O) + O_2 + PhNHNHPh \xrightarrow{Fe^{II}(MeCN)_4^{2+}} Ph_2S(O)_2 + H_2O + PhN{=}NPh$$

$$(6\text{-}16)$$

Table 6-4  Mono-oxygenation and Dehydrogenation of Organic Substrates via a
Model System for the Cytochrome *P*-450 Mono-oxygenase/Reductase Enzymes[a]

| Substrate (RH) | Product | % Reaction efficiency |
|---|---|---|
| $Ph_3P$ | $Ph_3PO$ | $100 \pm 5$ |
| $PhCH_2OH$ | $PhCH(O)$ | $30 \pm 4$ |
| $Ph_2SO$ | $Ph_2S(O)_2$ | $25 \pm 3$ |
| 1,4-cyclohexadiene | $PhH$ | $10 \pm 2$ |

[a] Solution conditions: 1 m$M$ $Fe^{II}(MeCN)_4^{2+}$, 1.6 m$M$ $O_2$, 1 m$M$ PhNHNHPh, 0.1 m$M$ $HO^-$, and 3 m$M$ substrate in MeCN. Reaction time, 0.5 h.

*Reaction    Mimic    for    the    Methane    Mono-oxygenase    Proteins;*
*$Fe^{II}(DPAH)_2/O_2/PhNHNHPh/substrate$*

The activation of dioxygen for the mono-oxygenation of saturated hydrocarbons
by the methane mono-oxygenase enzyme systems (MMO; hydroxylase/reductase)
represents an almost unique biochemical oxygenase, especially for the
transformation of methane to methanol.[4,5,16] The basic process involves the
insertion of an oxygen atom into the C–H bond of the hydrocarbon via the
concerted reduction of $O_2$ by the reductase cofactor

$$CH_4 + O_2 \xrightarrow{\text{MMO}/(H)_2} CH_3OH + H_2O \qquad (6\text{-}17)$$

An earlier section of this chapter (Table 6-2 and Scheme 6-1) describes
selective ketonization of methylenic carbons via activation of $O_2$ by $Fe^{II}(DPAH)_2$
to give $(DPAH)_2Fe^{III}OOFe^{III}(DPAH)_2$ as the reactive intermediate.

$$c\text{-}C_6H_{12} + O_2 \xrightarrow[\text{py}_2(\text{HOAc})]{Fe^{II}(DPAH)_2} c\text{-}C_6H_{11}OH + H_2O \qquad (6\text{-}18)$$

When an equimolar amount of $BrCCl_3$ (relative to $c$-$C_6H_{12}$) is present, $c$-$C_6H_{11}Br$
is the sole product (Table 6-5).[17] Replacement of the 1.8:1 py/HOAc solvent
system with MeCN or 3:1 MeCN/py causes the $Fe^{II}(DPAH)_2/O_2$ combination to
be unreactive with hydrocarbons, but it undergoes auto-oxidation to give
$(DPAH)_2Fe^{III}OFe^{III}(DPAH)_2$ (Scheme 6-1). However, the presence of excess
PhNHNHPh gives a system that is a hydrocarbon mono-oxygenase ($c$-$C_6H_{12} \rightarrow$
$c$-$C_6H_{11}OH$). The products and reaction efficiencies for various concentrations of
$Fe^{II}(DPAH)_2$ and PhNHNHPh with several substrates are summarized in Table
6-5.

The maximum efficiency and mono-oxygenase selectivity is achieved with
5 m$M$ $Fe^{II}(DPAH)_2$, 200 m$M$ PhNHNHPh, and 1 atm $O_2$ in 3:1 MeCN/py. The

distribution of $R$–OH isomers from 2-Me-butane indicates a selectivity in the order $\geq$CH >$\geq$CH$_2$ > –CH$_3$; the relative reactivities per C–H bond are 1.00, 0.29, and 0.05, respectively. With Fe$^{II}$(PA)$_2$/HOOH Fenton chemistry in 1.8:1 py/HOAc the relative reactivities are 1.00, 0.43, and 0.07,[18] and the values for aqueous HO· are 1.00, 0.48, and 0.10.[19] Thus, the reactive intermediate from the Fe$^{II}$(DPAH)$_2$/O$_2$/PhNHNHPh system is more selective than Fenton-derived and free HO·.

In the absence of an activating agent (BrCCl$_3$, BuI, or PhNHNHPh) the combination of Fe$^{II}$(DPAH)$_2$ and O$_2$ leads to the formation of (DPAH)$_2$Fe$^{III}$OFe$^{III}$(DPAH)$_2$ (4) via the transient formation of (DPAH)$_2$Fe$^{III}$OOFe$^{III}$(DPAH)$_2$ (3) (ketonizes methylenic carbons),[11] which requires that a 1:1 adduct [(DPAH)$_2$Fe(O$_2$) (2)] be formed initially via a rate-limiting step (Scheme 6-3). The presence of activating agents trap 2, especially when the [Fe$^{II}$(DPAH)$_2$]/[O$_2$] ratio is less than unity. This conclusion and the results of Table 6-5 are the basis for the proposed reaction pathways of Scheme 6-3.

The ability of the Fe$^{II}$(DPAH)$_2$/O$_2$/PhNHNHPh system (where PhNHNHPh is a mimic for flavin reductases)[13,15] to mono-oxygenate saturated hydrocarbons closely parallels the chemistry of the methane mono-oxygenase proteins.[4,5,16] However, the enzyme oxygenates 2-Me-butane with an isomer distribution of 82% primary alcohol, 10% secondary, and 8% tertiary.[20] The present model gives a distribution of 21% primary, 29% secondary, and 50% tertiary. Clearly the protein affords a cavity that is selective for CH$_4$ and –CH$_3$ groups. Although the likely reactive intermediates (7 and 8, Scheme 6-3) of the model are less reactive than free HO·, they are able to oxygenate –CH$_3$ groups and benzene (Table 6-5).

### Metal-surface-induced activation of dioxygen for oxygenase chemistry

*Ethylene oxygenation*

Supported silver is used for the activation of the O$_2$ in ambient air for the gas-phase selective epoxidation of ethylene.

$$2\,CH_2=CH_2 + O_2 \xrightarrow{\text{Ag}} 2\,H_2C{-}CH_2 \overset{\displaystyle \ulcorner O \urcorner}{} \tag{6-19}$$

Although the mechanism for O$_2$ activation is not known, the discussions in Chapter 4 indicate that the model for compound I of horseradish peroxidase, (Cl$_8$TPP)$^{+}$·Fe$^{IV}$=O, is an especially effective epoxidation reagent. The bond energy (-$\Delta G_{BF}$) for the Fe=O bond in this model is 46 kcal, which is in accord with its facile epoxidation of olefins (-$\Delta G_{BF}$, 77 kcal). Because the gas-phase bond energy ($\Delta H_{DBE}$) for the Ag=O bond is about 53 kcal (-$\Delta G_{BF}$ ~ 45 kcal),[21] this entity should be an effective O-atom transfer agent for olefin epoxidation.

Table 6-5  Activation of $O_2$ by $Fe^{II}(DPAH)_2$ via $BrCCl_3$ and PhNHNHPh for the Oxidation and Oxygenation of Hydrocarbons[a]

a.  1.8:1 py/HOAc; $O_2$ (1 atm, 3.4 mM); 10 h

| $Fe^{II}(DPAH)_2$ (mM) | Activator (M) | Substrate (1 M) | Products (mM)[b] |
|---|---|---|---|
| 32 | — | $c\text{-}C_6H_{12}$ | $c\text{-}C_6H_{10}(O)$ (4.4) |
|  |  | — | py(Br) (3.0) |
| 32 | $BrCCl_3$ (1.0) | $c\text{-}C_6H_{12}$ | $c\text{-}C_6H_{11}Br$ (3.2) |
| 32 | $BrCCl_3$ (1.0) | $c\text{-}C_6H_{12}$ | $c\text{-}C_6H_{11}Br$ (1.5), $c\text{-}C_6H_{10}(O)$ (1.8) |
| 32 | $BrCCl_3$ (0.01) | $c\text{-}C_6H_{12}$ | $c\text{-}C_6H_{11}Br$ (1.2) |
| 3 | $BrCCl_3$ (1.0) | $c\text{-}C_6H_{12}$ | $c\text{-}C_6H_{10}(O)$ (21.2) |
| 32 | PhNHNHPh (1.0) | $c\text{-}C_6H_{12}$ | $c\text{-}C_6H_{10}(O)$ (20.9) |
| 3 | PhNHNHPH (0.1) | $c\text{-}C_6H_{12}$ |  |

b.  3:1 MeCN/py; $O_2$ (1 atm, 7 mM); 22 h

| $Fe^{II}(DPAH)_2$ (mM) | Activator (M) | Substrate (1 M) | Products (mM)[b] |
|---|---|---|---|
| 28 | — | $c\text{-}C_6H_{12}$ | $(DPAH)_2Fe(III)OFe(III)(DPAH)_2$ (14) |
| 5 | PhNHNHPh (0.2) | $c\text{-}C_6H_{12}$[c] | $c\text{-}C_6H_{11}OH$ (35), $c\text{-}C_6H_{10}(O)$ (5.6) |
| 5 | PhNHNHPh (0.2) | $c\text{-}C_6H_{11}OH$ | $c\text{-}C_6H_{10}(O)$ (24) |
| 5 | PhNHNHPh (0.2) | $n\text{-}C_6H_{14}$ | $s\text{-}C_6H_{13}OH$ (19), $n\text{-}C_6H_{13}OH$ (1.5), $2\text{-}C_6H_{12}(O)$ (7.6) |

| | | | |
|---|---|---|---|
| 5 | PhNHNHPh (0.2) | Me$_2$CHCH$_2$Me | C$_5$H$_{11}$OH (17) [pr/s/t; 21:29:50],[d] Me$_2$CHC(O)Me (2.5) |
| 5 | PhNHNHPh (0.2) | PhCH$_2$Me | PhCH(OH)Me (1.5), PhC(O)Me (13) |
| 5 | PhNHNHPh (0.2) | PhCH$_3$ | MePhOH (3), PhCH(O) (3) |
| 5 | PhNHNHPh (0.2) | PhH | PhOH (3) |

[a] Substrate, activating agent, and Fe$^{II}$(DPAH)$_2$ [Fe$^{II}$(MeCN)$_4$(ClO$_4$)$_2$ added to 2 equivalents of (Me$_4$N)$_2$DPA] combined in 3.5 mL of solvent, followed by the addition of 1 atm of O$_2$ in a reaction cell with 18 mL of head space. Reaction temperature, 24±2°C.

[b] The product solutions were analyzed by capillary gas chromatography and GC-MS.

[c] Combination of 5 mM Fe$^{II}$(DPAH)$_2$, 1 M c-C$_6$H$_{12}$, and 100 mM HOOH in MeCN yields an ol/one product ratio of 2.3.

[d] Product profile for R· in a Fenton system; 2-Me-butane [25:35:40], Ref. 18.

Scheme 6-3  Activation of Dioxygen by $Fe^{II}(DPAH)_2$

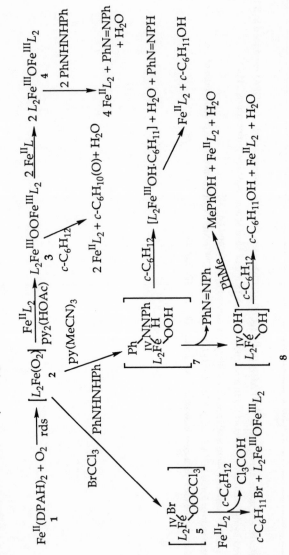

If the silver surface is an infinite oligomer of trivalent Ag atoms $(Ag^{III})_n$, then a reasonable $O_2$-activation reaction should lead to $(Ag^{III})_{n-1}$-$Ag^{III}$=O groups on the surface.

$$(Ag^{III})_n + O_2 \longrightarrow \left[ (Ag^{III}) \underset{n-2}{\overset{Ag^{III}}{\underset{Ag^{III}}{\diagdown\diagup}}} \begin{matrix} O \\ | \\ O \end{matrix} \right] \longrightarrow \left[ O=Ag^{III} \ (Ag^{III})_{\overline{n-2}} Ag^{III}=O \right]$$

$$(6\text{-}20)$$

Such groups will have Ag=O bond energies ($-\Delta G_{BF}$) that are less than 45 kcal, and should be effective epoxidation agents.

## Cyclohexane oxygenation

Another heterogeneous $O_2$-activation system uses supported cobalt metal to catalyze the gas-phase oxygenation of cyclohexane to cyclohexanol and cyclohexanone, which are intermediates for the production of adipic acid $[HOC(O)(CH_2)_4C(O)OH]$.

$$c\text{-}C_6H_{12}(g) + O_2(g) \xrightarrow{\ (Co)_n\ } c\text{-}C_6H_{11}OH + c\text{-}C_6H_{10}(O) + H_2O \qquad (6\text{-}21)$$

As in the case of silver, the cobalt surface probably is an infinite oligomer of trivalent cobalt atoms $(Co^{III})_n$ that may activate $O_2$ to give reactive intermediates

such as $(Co^{III})_{\overline{n-1}} Co^{III}$=O and $\underset{\underset{O-O}{\underbrace{\qquad}}}{Co^{III} (Co^{III})_{n-2} Co^{III}}$. The latter should have similar

reactivity to the intermediate from the $Co^{II}(bpy)_2^{2+}$/HOOH system (Chapter 4), which ketonizes methylenic carbons $[c\text{-}C_6H_{12} \longrightarrow c\text{-}C_6H_{10}(O) + H_2O]$. Thus, a reasonable reaction pathway is outlined by Eq. (6-22).

$$(Co^{III})_n(s) + O_2 \longrightarrow \left[ \underset{\underset{O-O}{\underbrace{\qquad}}}{Co^{III} (Co^{III})_{n-2} Co^{III}} \right]_{(s)} \xrightarrow{\ c\text{-}C_6H_{12}\ } \begin{matrix} c\text{-}C_6H_{10}(O) \\ \\ + H_2O + (Co^{III})_n \end{matrix}$$

$$(6\text{-}22)$$

Dioxygen appears to form similar $M-O-O-M$ groups on all transition-metal surfaces, but most systems rapidly degrade to $M-O-M$.

$$(M)_{\overline{n-1}}M-O-O-M-(M)_{n-1} \longrightarrow (M)_{\overline{n-2}}M-O-M-O-M-(M)_{n-1} \qquad (6\text{-}23)$$

This is especially true for manganese, iron, and coinage-metal (Cu, Ag, Au) surfaces,[22] but apparently is not favored with cobalt.

**Biological systems**

Nature uses a variety of metalloproteins and flavoproteins to activate (or catalyze) dioxygen for useful chemistry.[23-26] These can be subdivided into (1) oxidases (electron-transfer oxidation by $O_2$; e.g., cytochrome-*c* oxidase), (2) dehydrogenases (removal of two hydrogen atoms from the substrate by $O_2$; many so-called oxidases actually dehydrogenate substrates, e.g., glucose oxidase and xanthine oxidase), (3) mono-oxygenases [oxygen-atom transfer to substrate from $O_2$; usually have a reductase (hydrogenase) co-factor to reduce the second oxygen atom of $O_2$ to $H_2O$ (hence, the archaic name of "mixed-function" oxidases); e.g., cytochrome *P*-450 and methane mono-oxygenase], and (4) dioxygenases (transfer of the two oxygen atoms of $O_2$ to a substrate molecule, e.g., pyrocatechol dioxygenase). Each class of $O_2$-activating systems includes a variety of proteins and functions, and it is beyond the scope of this monograph to discuss the chemistry of individual $O_2$/enzyme/substrate processes. Instead, an attempt is made to bring to bear the chemistry of $O_2$ activation from the earlier sections of this chapter on a specific example for each class of $O_2$-activating proteins.

*Cytochrome-c oxidase*

Within aerobic biology the "harvesting" of the oxidative energy of $O_2$ is fundamental to oxidative metabolism. Because this is accomplished via the electron-transfer oxidation of four cytochrome *c*[Fe(II)] molecules ($E^{o'}_{pH\,7}$, +0.3 V versus NHE) per $O_2$, the challenge is to facilitate the reduction of $O_2$ via four one-electron steps, each with a potential greater than +0.4 V versus NHE at *p*H 7

$$O_2 + 4\,(H_3O^+)_{pH\,7} + 4\text{ cyt }c[Fe(II)] \xrightarrow{\text{cyt }c\text{ oxid}} 4\text{ cyt }c[Fe(III)]^+ + 6\,H_2O \quad (6\text{-}24)$$

The thermodynamics for the uncatalyzed electron-transfer reduction of $O_2$ does not meet this criterion (see Chapter 2),

$$O_2 \xrightarrow[-0.16\text{ V versus NHE}]{e^-} O_2^{-\cdot} \xrightarrow[+0.89\text{ V}]{2\,H_3O^+,\,e^-} HOOH$$
$$\xrightarrow[+0.40\text{ V}]{H_3O^+,\,e^-} HO\cdot \quad \xrightarrow[+2.31\text{ V}]{H_3O^+,\,e^-} 2\,H_2O \qquad (6\text{-}25)$$
$$2\,H_2O \qquad\qquad 2\,H_2O$$

especially for the initial *p*H-independent electron transfer to $O_2$.

The cytochrome-*c* oxidase protein is thought to consist of two heme-iron centers [heme *a* with two axial histidines and heme $a_3$ with one axial histidine (analogous to myoglobin)] and two copper centers [$Cu_A$ with two histidine, two cysteine, and one water/tyrosine ligand in its oxidized state; and $Cu_B$ with three histidine, one methionine, and one $H_2O/HO^-$ ligands].[27,28] The $Cu_A$/heme *a* pair constitute two coupled, one-electron redox couples (low potential, ~0.4 V) that facilitate (1) electron transfer from cytochrome $c$[Fe(II)] at the matrix side of the inner mitochondrial membrane as well as (2) proton transfer from the mitochondrial matrix across the inner membrane to the cytosol. At the cytosol side of the inner mitochondrial membrane the $Cu_B$/heme $a_3$ pair constitute the binding site for $O_2$ as well as the conduit for its high-potential four-electron, four-proton reduction to two $H_2O$ molecules.

Although recent discussions[27,28] of the mechanisms for $O_2$ reduction by cytochrome-$c$ oxidase involve metal-centered electron transfer, the chemistry of metal-catalyzed reductions usually involves metal-$O_2$ covalent bond formation and electron (or hydrogen-atom) transfer to the bound dioxygen (see Chapters 2 and 3). On this basis the mechanism for $O_2$ reduction by cytochrome-$c$ oxidase is reformulated in Scheme 6-4, with reducing equivalents transferred to the bound $O_2$ (believed to be the entity with the largest electron affinity, Chapter 3). This approach avoids the ionic formulations, which are inconsistent with metal/$O_2$ chemical processes, and achieves redox potentials that are greater than +0.6 V versus NHE for each of the four one-electron reductions.

**Scheme 6-4**    *Proposed mechanism for $O_2$ reduction via cytochrome c oxidase ($Cu_B$/heme $a_3$)*

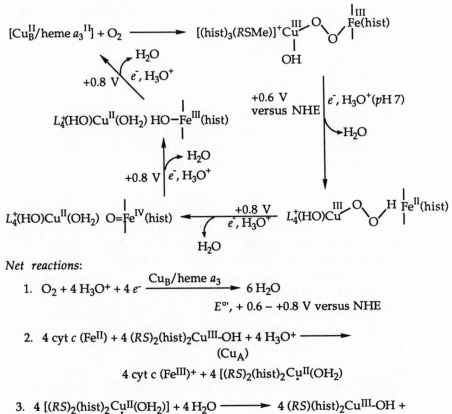

Net reactions:

1. $O_2 + 4\,H_3O^+ + 4\,e^- \xrightarrow{\;Cu_B/\text{heme }a_3\;} 6\,H_2O$

$$E^{\circ\prime},\ + 0.6 - +0.8 \text{ V versus NHE}$$

2. $4 \text{ cyt } c \text{ (Fe}^{II}) + 4\,(RS)_2(\text{hist})_2Cu^{III}\text{-OH} + 4\,H_3O^+ \longrightarrow$
$$(Cu_A)$$
$$4 \text{ cyt } c \text{ (Fe}^{III})^+ + 4\,[(RS)_2(\text{hist})_2Cu^{II}(OH_2)]$$

3. $4\,[(RS)_2(\text{hist})_2Cu^{II}(OH_2)] + 4\,H_2O \longrightarrow 4\,(RS)(\text{hist})_2Cu^{III}\text{-OH} +$
$$4\,H_3O^+ + 4\,e^-$$

*Mono-oxygenases*

Cytochrome $P$-450, which is the most extensively studied of the mono-oxygenase proteins, has a heme-iron active center with an axial thiol ligand (a cysteine residue). However, most chemical-model investigations[3,23] use simple iron(III) porphyrins without thiolate ligands. As a result, model mechanisms for cytochrome $P$-450 invoke a reactive intermediate that is formulated to be equivalent to Compound I of horseradish peroxidase, $(por^{+\cdot})Fe^{IV}=O$, with a high-potential porphyrin cation radical. Such a species would be reduced by thiolate, and therefore is an unreasonable formulation for the reactive center of cytochrome $P$-450.

Again, on the basis of (1) the chemistry for $O_2$ in the presence of transition metals and reducing agents (Chapter 2) and (2) reasonable bonding energetics for the components of the cytochrome $P$-450 mono-oxygenase/reductase system $[(por)Fe^{III}-SR/O_2/(NADH, H_3O^+/flavin)/substrate]$, a self-consistent mechanism is proposed and outlined in Scheme 6-5. The presence of an axial Fe–SR bond enhances the covalence of the iron, reduces the Fe–O bond energy, and thereby increases the radical character and reactivity of the $(RS-Fe^{IV}-O\cdot)$ group, and precludes formation of a high-potential porphyrin-cation-radical one-electron oxidant. Methane mono-oxygenase (MMO), like cytochrome $P$-450, activates $O_2$ via a reductase co-factor for the mono-oxygenation of saturated hydrocarbons (including $CH_4$ and $CH_3CH_3$).[4,5]

$$CH_4 + O_2 + 2\,(HA + e^-) \xrightarrow[\text{reductase}]{\text{MMO}} CH_3OH + H_2O + 2\,A^- \qquad (6\text{-}26)$$

The oxidized hydroxylase component of MMO includes a di-μ-hydroxo-bridged binuclear iron center. One of the iron atoms apparently is coordinately unsaturated,[4] and is thought to be the site for $O_2$ binding when both iron atoms have been reduced by the reductase co-factor of MMO.

In an earlier part of this chapter a system $[Fe^{II}(DPAH)_2/O_2/PhNHNHPh]$ has been discussed that mimics the catalytic cycle of the hydroxylase/reductase components of MMO (Scheme 6-3). On the basis of these results and the structural character of the hydroxylase component,[4,5] a reasonable mechanism for the mono-oxidation of saturated hydrocarbons by MMO is proposed and outlined in Scheme 6-6. Thus, Nature has devised a system to generate (via $O_2$ plus a reductase) two $HO\cdot$ radicals within the reaction complex, with one almost as reactive as free $HO\cdot$ and the other stabilized by an $[Fe^{IV}-OH]$ bond ($-\Delta G_{BF}$, ~40 kcal) such that it will only be reactive with the just-formed carbon radical ($R\cdot$).

The third example of a mono-oxygenase, $p$-hydroxybenzoate hydroxylase (HBH), makes use of a flavin cofactor in combination with a reductase.[29]

$$HO-\!\!\left\langle\!\bigcirc\!\right\rangle\!\!-C(O)OH + O_2 + NADPH + H_3O^+ \xrightarrow[H_2O]{HBH} HO-\!\!\left\langle\!\bigcirc\!\right\rangle\!\!-C(O)OH + H_2O + NADP^+$$

$$(6\text{-}27)$$

*Scheme 6-5 Proposed mechanism for the activation of $O_2$ for the
mono-oxygenation of substrates (RH) by cytochrome P-450*

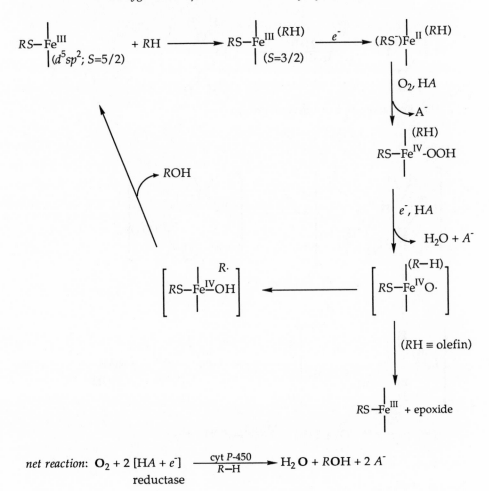

net reaction: $O_2 + 2\,[HA + e^-] \xrightarrow[\substack{cyt\ P\text{-}450 \\ R\text{-}H}]{} H_2O + ROH + 2\,A^-$
    reductase

Spectral studies of rapid-quench experiments indicate that the substrate/oxidized flavin/$O_2$/reductase combination forms an initial reactive intermediate that subsequently oxygenates the substrate.[30,31] These results in combination with the bonding arguments of Chapter 3 prompt the formulation in Scheme 6-7 of a general reaction mechanism for flavin mono-oxygenases. In the absence of substrate the system reduces $O_2$ to HOOH. In either case reduction is by H-atom transfer rather than by electron transfer.

*Scheme 6-6*    *Proposed mechanism for the activation of O₂ by methane mono-
oxygenase (MMO) for the monooxygenation of hydrocarbons (RH)*

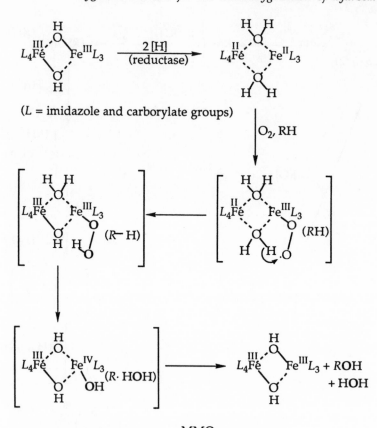

(L = imidazole and carborylate groups)

net reaction: $RH + O_2 + 2\,[H] \xrightarrow{\text{MMO}} ROH + H_2O$

## Dioxygenases

Pyrocatechase (PC) is representative of a class of proteins that facilitate the
transfer of both atoms of $O_2$ to a substrate molecule. With this particular
dioxygenase pyrocatechol is transformed to muconic acid.

$$\text{[pyrocatechol]} + O_2 \xrightarrow{\text{PC}} HOC(O)CH=CH\text{–}CH=CHC(O)OH \quad (6\text{-}28)$$

There is agreement that the enzyme contains nonheme iron at the active site
with N or O ligands (probably histidine, tyrosine, and/or asptartate residues),[32,33]
but less certainty as to the oxidation state, coordination sphere, and mode of $O_2$
binding and activation for the iron center. Earlier in this chapter the activation
of $O_2$ by $Fe^{II}(DPAH)_2$ for the transformation of pyrocatechol to muconic acid is
discussed (Table 6-2 and Scheme 6-1). Because this parallels the chemistry of the

*Scheme 6-7  Proposed mechanism for the activation of $O_2$ by flavin mono-oxygenase/NADH($H^+$) for the hydroxylation of aromatic substrates*

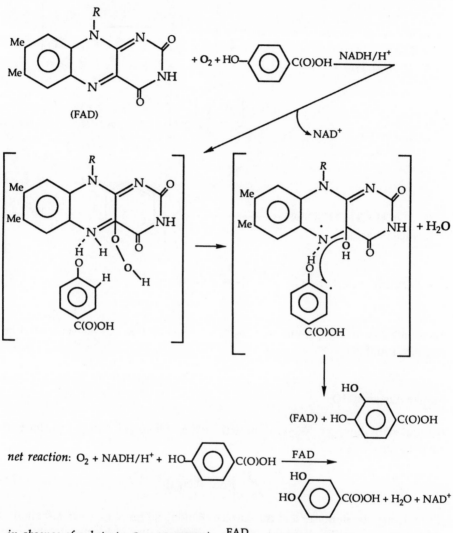

net reaction: $O_2$ + NADH/$H^+$ + HO—⟨◯⟩—C(O)OH $\xrightarrow{\text{FAD}}$ ... C(O)OH + $H_2O$ + NAD$^+$

in absence of substrate: $O_2$ + NADH/$H^+$ $\xrightarrow{\text{FAD}}$ HOOH + NAD$^+$

enzyme [Eq (6-28)], some of the essential elements of $O_2$ binding and activation may be the same. On the basis of what is known for the enzymatic process and the $Fe^{II}(DPAH)_2/O_2$/pyrocatechol system, a nonionic oxygen-atom-transfer mechanism for $O_2$ activation by pyrocatechol dioxygenase is proposed and outlined in Scheme 6-8. Key elements of the mechanism include recognition that (1) only reduced iron binds dioxygen, (2) the net process is two O-atom transfers per substrate and not electron transfer, and (3) only a single dioxygenated product is formed. The latter requires a reactive intermediate with

*Scheme 6-8*   *Proposed mechanism for the activation of $O_2$ by pyrocatechase (PC)*
*for the dioxygenation of catechols*

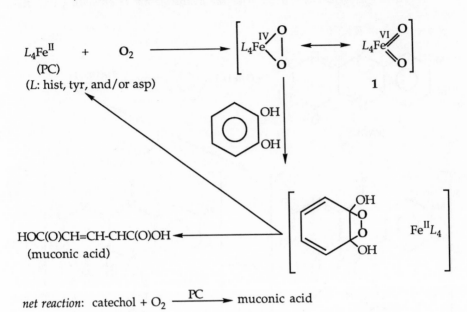

*net reaction:*  catechol + $O_2$ $\xrightarrow{\text{PC}}$ muconic acid

two equivalent oxygen atoms (species **1**, Scheme 6-8) that can be transferred in a concerted manner.

## Singlet dioxygen ($^1O_2$)

Ground-state dioxygen ($^3O_2$) can be activated by photolytic energy transfer to the singlet state ($^1O_2$),

$$^3O_2 \xrightarrow[\text{Rose Bengal}]{h\upsilon} {}^1O_2 \qquad (6\text{-}29)$$

and this may be duplicated at the surface of biomembranes by solar radiation. In the laboratory singlet dioxygen is conveniently produced by the stoichiometric oxidation of hydrogen peroxide by hypochlorous acid.

$$HOOH + HOCl \longrightarrow {}^1O_2 + H_2O + HCl \qquad (6\text{-}30)$$

Although there is considerable emotion with respect to the hazardous reactivity of singlet dioxygen in biological systems, its chemistry is highly selective; $^1O_2$ is unreactive with most organic functional groups (e.g., saturated σ bonds, simple olefins, and carbonyl groups). In part, this is due to its limited lifetime (2 μs in $H_2O$, 58 μs in $^2H_2O$, 16 μs in DMF, 65 μs in MeCN, and 610 μs in $d_3$-MeCN),[34,35] and to its paired valence electron shell and to its strong O–O bond

($\Delta H_{DBE}$, 119 kcal). The suspected biological hazard of $^1O_2$ has prompted extensive research with a substantial literature in several disciplines. A recent review[36] provides a comprehensive overview of the reactivity for $^1O_2$ in various phases and matrices. The most facile chemistry for $^1O_2$ is addition to conjugated polymers and polynuclear aromatic molecules.[37,38]

$$CH_2=CH-CH=CH_2 + {}^1O_2 \longrightarrow \quad \overset{\displaystyle OO}{\underset{\displaystyle CH=CH}{CH_2 \quad CH_2}} \qquad (6\text{-}31)$$

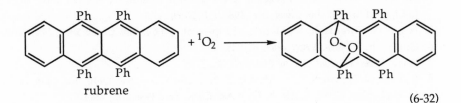

rubrene

(6-32)

1,3-diphenylisobenzofuran (DPBF)   *o*-dibenzoylbenzene (DBB)

(6-33)

Unfortunately, much of the chemistry attributed to $^1O_2$ actually is due to a precursor in its generation cycle. Thus, many have used cis-stilbene (*c*-PhCH=CHPh) to indicate the presence of $^1O_2$ via the production of PhCH(O).[39] Careful experiments with authentic photochemically generated $^1O_2$ [Eq. (6-23)] have established that *c*-PhCH=CHPh has a small reaction rate.[40] Hence, the large yields of PhCH(O) that are attributed to $^1O_2$ actually result from oxygenation by an $^1O_2$-precursor molecule.

A specific example from Chapter 4 will illustrate the point. In pyridine/acetic acid the combination of Fe(II)(PA)$_2$ and HOOH yields a reactive intermediate that dioxygenates *c*-PhCH=CHPh

$$[(PA)_2Fe^{III}OOFe^{III}(PA)_2] + c\text{-PhCH=CHPh} \xrightarrow{\text{py}_2(HOAc)} 2\,PhCH(O) + 2\,Fe^{II}(PA)_2 \quad (6\text{-}34)$$
$$\xrightarrow{\text{DMF}} {}^1O_2 + 2\,Fe^{II}(PA)_2$$

When this experiment is done in DMF *c*-PhCH=CHPh is not reactive. The reactive intermediate apparently is formed, but its lifetime is too short and it decomposes to give a stoichiometric yield of $^1O_2$.[12] Likewise, the reactive intermediate from the combination of $Fe^{II}(MeCN)_4^{2+}$ and excess HOOH in MeCN,

$(MeCN)_4^{2+} Fe^{IV} \overset{O}{\underset{O}{\diagdown}}$ ' reacts with $c$-PhCH=CHPh (52% efficiency) to give two PhCH(O) molecules, but does not produce measurable quantities of $^1O_2$ (Scheme 4-2, Chapter 4).[39]

## References

1. Olafson, B. D.; Goddard, W. A., III, *Proc. Natl. Acad. Sci. USA* **1977**, *74*, 1315.
2. Collman, J. P. ; Gagné, R. R.; Reed, C. A.; Halbert, T. R.; Lang, G.; Robinson, W. T. *J. Am. Chem. Soc.* **1975**, *97*, 1427.
3. Ortiz de Montellado, P. R. (ed). *Cytochrome P-450*. New York: Plenum, 1985.
4. Ericson, A.; Hedman, B.; Hodgson, K. O.; Green J.; Dalton, H.; Bentsen, J. G.; Beer, S. J.; Lippard, S. J. *J. Am. Chem. Soc.* **1988**, *110*, 2330.
5. Fox, B. G.; Surerus, K. K.; Munck, E.; Lipscomb, J. D. *J. Biol. Chem.* **1988**, *263*, 10553.
6. Que, L., Jr. *J. Chem. Educ.* **1985**, *62*, 938.
7. Foote, C. S. *Free Radicals in Biology* **1976**, *2*, 85.
8. (a) Chin, D.-H.; Balch, A. L.; MaMar, G. N. *J. Am. Chem. Soc.* **1990**, *102*, 1446. (b) Chin, D.-H.; LaMar, G. N.; Balch, A. L.; *J. Am. Chem. Soc.* **1980**, *102*, 4344.
9. Simic , M. G.; Karel, M. (eds.). *Autoxidation in Food and Biological Systems*. New York: Plenum, 1980.
10. Sobkowiak, A.; Sawyer, D. T. *Inorg. Chem.*, submitted, September, 1990.
11. Sheu, C.; Sobkowiak, A.; Jeon, S.; Sawyer, D. T. *J. Am. Chem. Soc.* **1990**, *112*, 879.
12. Sheu, C.; Richert, S. A.; Cofré, P.; Ross, B., Jr.; Sobkowiak, A.; Sawyer, D. T.; Kanofsky, Jr. *J. Am. Chem. Soc.* **1990**, *112*, 1936.
13. Calderwood, T. S.; Johlman, C. L.; Roberts, J. L., Jr., Wilkins, C. L.; Sawyer, D. T. *J. Am. Chem. Soc.* **1984**, *106*, 4683.
14. Weller, M. G.; Weser, J. *J. Am. Chem. Soc.* **1982**, *104*, 3752.
15. Sawyer, D. T.; Sugimoto, H.; Calderwood, T. S. *Proc. Natl. Acad. Sci., USA* **1981**, *81*, 8025.
16 Dalton, H. *Adv. Appl. Microbiol.* **1980**, *26*, 71.
17. Sheu, C.; Sawyer, D. T.; *J. Am. Chem. Soc.* **1990**, *112*, 8212.
18. Sheu, C.; Sobkowiak, A.; Zhang, L.; Ozbalik, N.; Barton, D. H. R.; Sawyer, D. T. *J. Am. Chem. Soc.* **1989**, *111*, 8030.
19. Trotman-Dickenson, A. F. *Adv. Free Radical Chem.* **1965**, *1*, 1.
20. Lipscomb, J. D., University of Minnesota, private communication; March 30, 1990.
21. Lide, D. R. (ed.). *CRC Handbook of Chemistry and Physics*, 71st ed. Boca Raton, FL: CRC, 1990, pp. 9-86–98.
22. Sawyer, D. T.; Chooto, P.; Tsang, P. K. S. *Langmuir* **1989**, *5*, 84.
23. King, T. E.; Mason, H. S.; Morrison, M. (eds.). *Oxidases and Related Redox Systems*. New York: Alan R. Liss, Inc., 1988.
24. (a) Hazaishi, O. (ed.). *Oxygenases*. New York: Academic Press, 1962; (b) Hazaishi, O. (ed.). *Molecular Mechanisms for Oxygen Activation*. New York: Academic Press, 1974.
25. Bannister, J. V.; Bannister, W. H. (eds.). *The Biology and Chemistry of Active Oxygen*. New York: Elsevier, 1974.

26. Ingraham, L. L.; Meyer, D. L. *Biochemistry of Dioxygen*. New York: Plenum Press, 1985.

27. Chan, S. I.; Witt, S. N.; Blair, D. F. *Chemica Scripta* **1988**, *28A*, 51.

28. Chan, S. I.; Li, P. M.; Nilson, T.; Blair, D. F.; Martin, C. T. In *Oxidases and Related Redox Systems* (King, T. E.; Mason, H. S.; Morrison, M., eds.). New York: Alan R. Liss, Inc., 1988, pp. 731–747.

29. Massey, V.; Hemmerick, P. In *The Enzymes, Vol. XII*; (Boyer, P., ed.). New York: Academic Press, 1975, pp. 191–252.

30. Entsch, B.; Massey, V.; Ballou, D. P. *Biochem. Biophys. Res. Commun.* **1974**, *57*, 1018.

31. Massey, V.; Schopfer, L. M.; Anderson, R. F. In *Oxidases and Related Redox Systems* (King, T. E.; Mason, H. S.; Morrison, M., eds.). New York: Alan R. Liss, Inc., 1988, pp. 147–166.

32. Que, L., Jr. *Structure and Bonding* **1980**, *40*, 39.

33. Nozaki, M.; Hazaishi, O. In *The Biology and Chemistry of Active Oxygen* (Bannister, J. V.; Bannister, W. H., eds.). New York: Elsevier, 1984, pp. 68–104.

34. Sugimoto, H.; Sawyer, D. T.; Kanofsky, J. R. *J. Am. Chem. Soc.* **1988**, *110*, 8707.

35. Ogilly, P. R.; Foote, C. S. *J. Am. Chem. Soc.* **1983**, *105*, 3423.

36. Frimer, A. A. (ed.). *Singlet O$_2$. Volume I. Physical-Chemistry Aspects. Volume II. Reaction Modes and Products, Part 1. Volume III. Reaction Modes and Products, Part 2. Volume IV. Polymers and Biomolecules*. Boca Raton, FL: CRC Press, 1985.

37. Foote, C. S. In *Biochemical and Clinical Aspects of Oxygen* (Caughey, W. S., ed.) New York: Academic Press, 1979; pp. 603–626.

38. Merkel, P. B.; Kearns, D. R. *J. Am. Chem. Soc.* **1972**, *94*, 7244.

39. Sugimoto, H.; Sawyer, D. T. *J. Am. Chem. Soc.* **1984**, *106*, 4283.

40. Foote, C. S.; University of California, Los Angeles, private communication.

# 7

# REACTIVITY OF SUPEROXIDE ION

*How super is superoxide?*

Reduction of dioxygen by electron transfer yields superoxide ion ($O_2^{-\cdot}$),[1] which has its negative charge and electronic spin density delocalized between the two oxygens. As such it has limited radical character [H–OO$^-$ bond energy $\Delta G_{BF}$, 72 kcal][2] and is a weak Brønsted base in water[3]

$$HOO\cdot \longrightarrow H^+ + O_2^{-\cdot} \qquad k_{diss}, 2.0 \times 10^{-5} \qquad (7\text{-}1)$$

The dynamics for the hydrolysis and disproportionation of $O_2^{-\cdot}$ in aqueous solutions have been characterized by pulse radiolysis.[3-7] For all conditions the rate-limiting step is second order in $O_2^{-\cdot}$ concentration, and the maximum rate occurs at a pH that is equivalent to the $pK_a$ for HOO$\cdot$ (it decreases monotonically with further decreases in the hydrogen ion concentration).

$$HOO\cdot + O_2^{-\cdot} \xrightarrow{H_3O^+} HOOH + O_2, k_2, 1.0 \times 10^8 \ M^{-1}s^{-1} \qquad (7\text{-}2)$$

$$HOO\cdot + HOO\cdot \longrightarrow HOOH + O_2, k_2, 8.6 \times 10^5 \ M^{-1}s^{-1} \qquad (7\text{-}3)$$

$$O_2^{-\cdot} + O_2^{-\cdot} + 2\,H_2O \longrightarrow HOOH + O_2 + 2\ ^-OH, k_4, <0.3 \ M^{-1}s^{-1} \qquad (7\text{-}4)$$

## Brønsted base

In water at $p$H 7 the disproportionation equilibruim is far to the right (Chapter 2)[8]

$$2\,O_2^{-\cdot} + 2\,H_3O^+ \rightleftharpoons HOOH + O_2, K, 4 \times 10^{20} \qquad (7\text{-}5)$$

and even at pH 14 it is complete

$$2\,O_2^{-\cdot} + H_2O \rightleftharpoons HOO^- + \,^-OH + O_2, K, 9.1 \times 10^8 \qquad (7\text{-}6)$$

Thus, the dominant characteristic of $O_2^{-\cdot}$ in any medium is its ability to act as a strong Brønsted base via formation of HOO$\cdot$,[9,10] which reacts with allylic hydrogens, itself, or a second $O_2^{-\cdot}$ (Scheme 7-1). Such a proton-drive disproportionation process means that $O_2^{-\cdot}$ can deprotonate acids much weaker than water (up to pKa $\approx$ 23).[11]

*Scheme 7-1 Proton-induced activation of $O_2^{-\cdot}$ in $Me_2SO$*

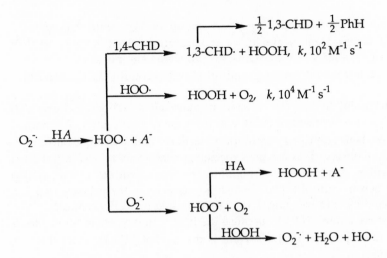

The propensity of $O_2^{-\cdot}$ to remove protons from substrates accounts for its reactivity with acidic reductants and their overall oxidation. Thus, combination of $O_2^{-\cdot}$ with protic substrates [α-tocopherol, hydroquinone, 3,5-di-*tert*-butylcatechol, L(+)-ascorbic acid] yields products that are consistent with an apparent one-electron oxidation of the substrate and the production of HOOH. However, the results of electrochemical studies[12] provide clear evidence that these substrates are not oxidized in aprotic media by direct one-electron transfer to $O_2^{-\cdot}$. The primary step involves abstraction of a proton from the substrate by $O_2^{-\cdot}$ to give substrate anion and the disproportionation products of HOO· (HOOH and $O_2$). In turn, the substrate anion is oxidized by $O_2$ in a multistep process to yield oxidation products and HOOH. Thus, by continuously purging the $O_2$ that results from the disproportionation of $O_2^{-\cdot}$ when it is combined with α-tocopherol (by vigorous Ar bubbling through the solution), quantitative yields of substrate anion are obtained without significant oxidation.

## Nucleophilicity

Although superoxide ion is a powerful nucleophile in aprotic solvents, it does not exhibit such reactivity in water, presumably because of its strong solvation by that medium ($\Delta H_{hydration}$, 100 kcal) and its rapid hydrolysis and disproportionation. The reactivity of $O_2^{-\cdot}$ with alkyl halides via nucleophilic substitution was first reported in 1970.[13,14] These and subsequent kinetic studies[15-18] confirm that the reaction is first order in substrate, that the rates follow the order primary>secondary>>tertiary for alkyl halides and tosylates, and that the attack by $O_2^{-\cdot}$ results in inversion of configuration ($S_N2$).

*Oxygenation of halogenated hydrocarbons*

The stoichiometries and kinetics for the reaction of $O_2^{-\cdot}$ with halogenated hydrocarbons (alkanes, alkenes, and aromatics) are summarized in Table 7-1.[18-24] [The normalized first-order rate constants, $k_1/[S]$, were determined by the rotated ring-disk electrode method under pseudo-first-order conditions ([substrate] > $[O_2^{-\cdot}]$).[19]

The nucleophilicity of $O_2^{-\cdot}$ toward primary alkyl halides (Scheme 7-2) results in an $S_N2$ displacement of halide ion from the carbon center. The normal reactivity order, benzyl>primary>secondary>tertiary, and leaving-group order, I>Br>OTs>Cl, are observed, as are the expected stereoselectivity and inversion at the carbon center. In dimethylformamide the final product is the dialkyl peroxide. The peroxy radical (ROO·), which is produced in the primary step and has been detected by spin trapping,[25] is an oxidant that is readily reduced by $O_2^{-\cdot}$ to form the peroxy anion (ROO⁻). Because the latter can oxygenate $Me_2SO$ to its sulphone, the main product in this solvent is the alcohol (ROH) rather than the dialkyl peroxide.

*Scheme 7-2   Nucleophilic displacement of alkyl halides by $O_2^{-\cdot}$*

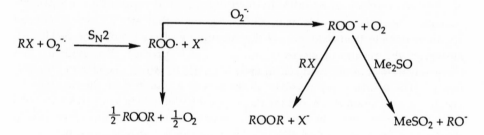

Although formation of the dialkyl peroxide is shown in the prototype reaction (Scheme 7-1), hydroperoxides, alcohols, aldehydes, and acids also have been isolated. The extent of these secondary paths depends on the choice of solvent and reaction conditions. Secondary and tertiary halides also give substantial quantities of alkenes from dehydrohalogenation by $O_2^{-\cdot}/HOO^-/HO^-$.

The reaction of $O_2^{-\cdot}$ with $CCl_4$ and $RCCl_3$ compounds almost certainly cannot occur via an $S_N2$ mechanism because the carbon-atom center is inaccessible. Rather, superoxide ion appears to attack a chlorine atom with a net result that is equivalent to an electron transfer from $O_2^{-\cdot}$ to chlorine (Scheme 7-3). This step is analogous to the "single-electron-transfer" (SET) mechanism that has been proposed for many nucleophilic reactions; an initial transfer of an electron followed by the collapse of a radical pair.[26]

$$E^+ + Nu^- \rightarrow [E \cdot \cdot Nu] \rightarrow E\text{--}Nu \tag{7-7}$$

Table 7-1 Stoichiometries, Products, and Apparent Second-Order Rate Constants ($k_1/[S]$) for the Reaction of Excess $O_2^-$ with Halogenated Hydrocarbons in Dimethylformamide (0.1 M Tetraethylammonium Perchlorate) at 25°C$^a$

| Substrate concentration, [S] = 1–10 mM | $[O_2^-]/[S]$ | [Products]/[S] | $k_1/[S]^b$ ($M^{-1} s^{-1}$) |
|---|---|---|---|
| a. Halo-alkanes | | | |
| MeCl | 1 | Cl⁻, 1/2 MeOOMe | $(8.0\pm1.0) \times 10^1$ |
| MeBr | 1 | Br⁻, 1/2 MeOOMe | $(4.8\pm1.0) \times 10^2$ |
| n-BuCl | 1 | Cl⁻, 1/2 BuOOBu | $(3.2\pm2) \times 10^0$ |
| n-BuBr | 1 | Br⁻, 1/2 BuOOBu | $(7.4\pm2.0) \times 10^2$ |
| s-BuBr | 1 | Br⁻, 1/2 BuOOBu | $(4.0\pm1.0) \times 10^2$ |
| | | | |
| b. Polyhalo-alkanes | | | |
| $CCl_4$ | 5 | HOC(O)O⁻, 4 Cl⁻, 3.3 $O_2$ | $(3.8\pm1.0) \times 10^3$ |
| $CBr_4$ | 5 | HOC(O)O⁻, 4 Br⁻, 3.3 $O_2$ | $>1 \times 10^5$ |
| $FCCl_3$ | 5 | HOC(O)O⁻, 3 Cl⁻, 2.5 $O_2$ | $(4.0\pm1.0) \times 10^0$ |
| $HCCl_3$ | 4 | HOC(O)O⁻, 3 Cl⁻, 2.0 $O_2$ | $(0.4\pm0.2) \times 10^0$ |
| $CF_3CCl_3$ | 4 | $CF_3$C(O)O⁻, 3 Cl⁻, 2.4 $O_2$ | $(4.0\pm1.0) \times 10^2$ |
| $PhCCl_3$ | 4 | PhC(O)O⁻ (70%), PhC(O)OO⁻ (30%), 3 Cl⁻, 2.4 $O_2$ | $(5.0\pm1.0) \times 10^1$ |
| $MeCCl_3$ | — | | <0.1 |
| $HOCH_2CCl_3$ | 4 | $HOCH_2$C(O)O⁻, 3 Cl⁻ | $(4.7\pm1.0) \times 10^1$ |
| $(p\text{-}ClPh)_2CHCCl_3$ (DDT) | 1 | $(p\text{-}ClPh)_2$C=$CCl_2$, Cl⁻ | $(1.0\pm0.2) \times 10^2$ |

| | | | |
|---|---|---|---|
| (p-MeOPh)$_2$CHCCl$_3$ (methoxychlor) | 1 | (p-MeOPh)$_2$C=CCl$_2$, Cl$^-$ | (1.0±0.2) × 10$^1$ |
| (p-ClPh)$_2$CFCCl$_3$ (F-DDT) | 1 | (p-ClPh)$_2$C=CCl$_2$, Cl$^-$ | (1.7±0.2) × 10$^2$ |
| H$_2$CCl$_2$ | 2 | H$_2$C(O), 2 Cl$^-$, 1.5 O$_2$ | (7.2±2.0) × 10$^0$ |
| H$_2$CBr$_2$ | 2 | H$_2$C(O), 2 Br$^-$, 1.5 O$_2$ | (2.3±0.5) × 10$^2$ |
| PhCHBrCHBrPh | 2 | 2 PhCH(O), 2 Br$^-$, O$_2$ | (1.0±0.2) × 10$^3$ |
| MeCHBrCHBrMe | 2 | 2 MeCH(O), 2 Br$^-$, O$_2$ | (1.6±0.2) × 10$^2$ |
| CH$_2$BrCH$_2$Br (EDB) | 2 | 2 CH$_2$ (O), 2 Br$^-$, O$_2$ | (2.0±0.4) × 10$^3$ |
| CH$_2$BrCHBrCH$_2$Cl (DBCP) | 5 | 2 CH$_2$(O), HOC(O)O$^-$, 2 Br$^-$, Cl$^-$, 2 O$_2$ | (4.0±1.0) × 10$^3$ |
| c. Halogenofluorocarbons | | | |
| F$_3$CBr | 2 | Br$^-$ | (2.6±1.0) × 10$^1$ |
| F$_2$CBr$_2$ | 5 | 2 Br$^-$ | (1.1±0.3) × 10$^3$ |
| FCCl$_3$ | 5 | 3 Cl$^-$ | (4.2±1.0) × 10$^0$ |
| FCBr$_3$ | 5 | 3 Br$^-$ | (2.2±0.9) × 10$^3$ |
| F$_3$CCCl$_3$ | 4 | 3 Cl$^-$ | (4.0±1.0) × 10$^2$ |
| d. Chloro-alkenes | | | |
| cis-CHCl=CHCl | 4 | 2 HOC(O)O$^-$, 2 Cl$^-$ | (1.0±0.3) × 10$^1$ |
| CH$_2$=CCl$_2$ | 3 | HOC(O)O$^-$, 2 Cl$^-$, O$_2$ | (2.0±0.5) × 10$^0$ |
| CHCl=CCl$_2$ | 5 | 2 HOC(O)O$^-$, 3 Cl$^-$, 1.5 O$_2$ | (9.0±1.0) × 10$^0$ |
| CCl$_2$=CCl$_2$ | 6 | 2 HOC(O)O$^-$, 4 Cl$^-$, 3O$_2$ | (1.5±0.4) × 10$^1$ |
| (p-ClPh)$_2$C=CCl$_2$ (DDE) | 3 | HOC(O)O$^-$, (p-ClPh)$_2$C=O, 2 Cl$^-$, O$_2$ | (2.0±0.5) × 10$^0$ |

e. Polychloro and polyfluoro aromatics

| | | | |
|---|---|---|---|
| $C_6Cl_6$ | 12 | 6 HOC(O)O$^-$, 6 Cl$^-$, 4.5 $O_2$ | $(1.0\pm0.3) \times 10^3$ |
| $HC_6Cl_5$ | 11 | 6 HOC(O)O$^-$, 5 Cl$^-$, 3 $O_2$ | $(8.0\pm2.0) \times 10^1$ |
| $H_2C_6Cl_4$ | 10 | 6 HOC(O)O$^-$, 4 Cl$^-$, 2 $O_2$ | $(2.0\pm0.5) \times 10^0$ |
| $Cl_5C_6-C_6Cl_5$ (PCB) | 22 | 12 HOC(O)O$^-$, 10 Cl$^-$, 7 $O_2$ | $(2.0\pm0.5) \times 10^2$ |
| $C_6F_6$ | 2 | $F_5C_6OO^-$, F$^-$, $O_2$ | $(3.0\pm0.6) \times 10^1$ |
| $HC_6F_5$ | 2 | $HC_6F_4OO^-$, F$^-$, $O_2$ | $(1.0\pm0.2) \times 10^1$ |
| $CF_3C_6F_5$ | 2 | $CF_3C_6F_4OO^-$, F$^-$, $O_2$ | $(1.0\pm0.2) \times 10^2$ |
| $F_5C_6-C_6F_5$ | 4 | $C_{12}F_8(OO)_2^{2-}$, 2 F$^-$, 2 $O_2$ | $(1.0\pm0.2) \times 10^2$ |

[a] Stoichiometries determined (a) for $O_2^{-\cdot}$/substrate reactions by titration of excess (Me$_4$N)O$_2$ (with voltammetric detection); (b) for released Br$^-$ and Cl$^-$ by titration with AgNO$_3$; for released $O_2$ by negative-scan voltammetry; and (c) for organic products by ether extractions and capillary-column gas chromatography.

[b] Pseudo-first-order rate constants, $k_1$ (normalized to unit substrate concentration [S]), were determined from measurements with a glassy-carbon–glassy-carbon ring-disk electrode that was rotated at 900 rev min$^{-1}$ (Ref. 19), or from the ratio ($i_{anodic}/i_{cathodic}$) for the cyclic voltammogram of $O_2$ in the presence of excess substrate (Ref. 20).

*Scheme 7-3   Nucleophilic degradation of carbon tetrachloride by $O_2^{-\cdot}$*

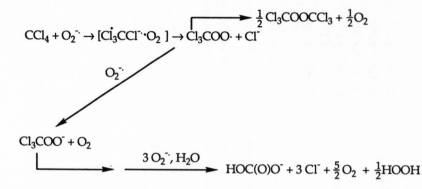

The initiation step for the $(O_2^{-\cdot})$-$CCl_4$ reactions must be followed by rapid combination in the solvent cage of $\cdot O_2\cdot$ and $Cl_3C\cdot$ to form the $Cl_3COO\cdot$ radical. This radical is thought to initiate lipid peroxidation,[27] which could account for the hepatoxicity of $CCl_4$.[28]

The rates of reaction for $O_2^{-}$ with $RCCl_3$ compounds are proportional to their reduction potentials, which is consistent with the SET mechanism.[19] A plot of log $k_1/[S]$ (Table 7-1) against the reduction potentials of $RCCl_3$ compounds is approximately linear with a slope of -4.9 decade per volt. Such behavior is consistent with a mechanism that occurs via simultaneous electron transfer and nuclear motion.[29] This correlation indicates that in water, where the $O_2/O_2^{-\cdot}$ redox potential is about 0.44 V more positive, the rate of the reaction for $O_2^{-\cdot}$ with $CCl_4$ would be about 100-200 times slower than in aprotic solvents.

The stoichiometric data in Table 7-1 for $CCl_4$ and $FCCl_3$ are consistent with a net chemical reaction that yields bicarbonate ion and four halide ions in the final aqueous workup of the reaction products. When $O_2^{-\cdot}$ reacts with $RCCl_3$ compounds, the R–C bond is not cleaved (Scheme 7-4).

*Scheme 7-4   Nucleophilic degradation of tri-chloro-methanes by $O_2^{-}$*

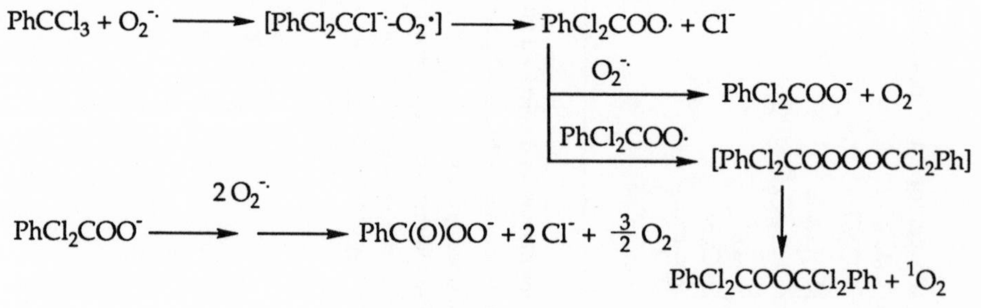

$$PhCCl_3 + O_2^{-\cdot} \longrightarrow [PhCl_2CCl^{-}\text{-}O_2{}^{\cdot}] \longrightarrow PhCl_2COO\cdot + Cl^{-}$$

$$\xrightarrow{O_2^{-\cdot}} PhCl_2COO^{-} + O_2$$

$$\xrightarrow{PhCl_2COO\cdot} [PhCl_2COOOOCCl_2Ph]$$

$$PhCl_2COO^{-} \xrightarrow{2\,O_2^{-\cdot}} \longrightarrow PhC(O)OO^{-} + 2\,Cl^{-} + \tfrac{3}{2}O_2$$

$$\downarrow$$

$$PhCl_2COOCCl_2Ph + {}^1O_2$$

About 30% of the product from the $PhCCl_3/O_2^{-\cdot}$ reaction is perbenzoate ion. Hence, this may be the active intermediate responsible for the epoxidation of olefins by $O_2^{-\cdot}$ in the presence of benzoyl chloride.[30] Further support for the formation of $PhCCl_2OO\cdot$ form the primary step of the $PhCCl_3/O_2^{-\cdot}$ reaction is provided by the observation of $^1O_2$ when a large excess of substrate is present (Scheme 7-4).[31,32]

Superoxide ion reacts with vicinal dibromoalkanes to form aldehydes (Table 7-1). The mechanism proposed for these reactions (Scheme 7-5) is a nucleophilic attack on carbon, followed by a one-electron reduction of the peroxy radical and nucleophilic displacement on the adjacent carbon to form a dioxetane that subsequently cleaves to form two moles of aldehyde.[21]

*Scheme 7-5   Nucleophilic degradation of α-dihalides by $O_2^{-\cdot}$*

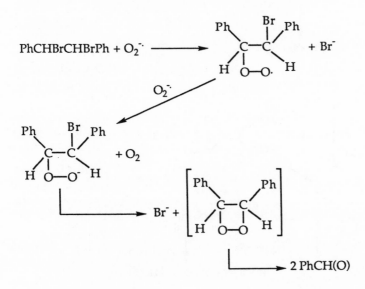

Both *p,p'*-DDT and methoxychlor are rapidly deprotonated by $HO^-$ in aprotic solvents with subsequent elimination of $Cl^-$ to form the dehydro–chlorination products; that is, DDT forms DDE (Scheme 7-6). The same products are formed in their reactions with $O_2^{-\cdot}$. Because the reaction rates that are measured by the rotated ring-disk electrode method are fairly rapid, the primary step must be a direct reaction with $O_2^{-\cdot}$ and not with $HO^-$ [from the reaction of $O_2^{-\cdot}$ with trace water in the solvent ($2\,O_2^{-\cdot} + H_2O \rightarrow O_2 + HOO^- + HO^-$)]. Hence, the initial reaction with $O_2^{-\cdot}$ is deprotonation followed by elimination of $Cl^-$ to form DDE.

The reactivities of chloro- and bromofluoromethanes (Freons and others) with superoxide ion are summarized in Table 7-1. Substitution of fluorine atoms into chloromethanes results in a substantial decrease in their reactivity with $O_2^{-\cdot}$ (relative rates: $CCl_4 >> CF_4$, $CCl_4 >> FCCl_3$, $CCl_4 >> F_2CCl_2$, $CCl_4 >> F_3CCl$, $H_3CCl >> F_3CCl$, and $H_2CCl_2 >> F_2CCl_2$). The bromo derivatives react much faster than the corresponding chloro compounds ($F_3CBr >> F_3CCl$, $F_2CBr_2 >> F_2CCl_2$, and

*Scheme 7-6  Nucleophilic degradation of DDT by $O_2^{-\cdot}$*

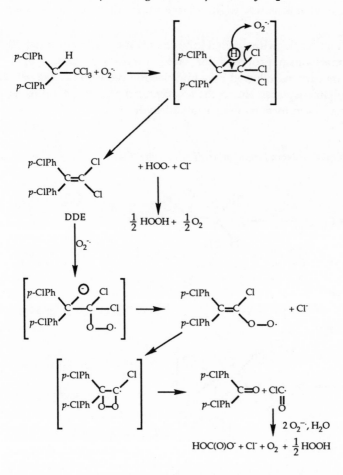

FCBr$_3$>>FCCl$_3$).  The rates of reaction for HCCl$_3$, HCFCl$_2$, and HCF$_2$Cl are approximately the same, and the reactions appear to have a common path via dehydrochlorination.  The reaction pathways of Schemes 7-2 and 7-3 appear to be followed by the halogenofluorocarbons.[22]

The overall reaction and product stoichiometries for the degradation of chloroalkene substrates by $O_2^{-\cdot}$ in DMF are summarized in Table 7-1.[20]  Within the limits of a reaction time of 10 min or less, chloroethene, *trans*-1,2-dichloroethene, Aldrin, and Dieldrin are not oxidized by $O_2^{-\cdot}$ in DMF.  A reasonable mechanism for these oxidations is an initial nucleophilic addition of superoxide to the chloroalkenes [e.g., tetrachloroethene (Scheme (7-7)].  Subsequent loss of chloride ion would give a vinyl peroxy radical, which can cyclize and decompose to a chloroacyl radical and phosgene.[31]  These would undergo subsequent facile reactions with $O_2^{-\cdot}$ to give bicarbonate and chloride ions.

*Scheme 7-7   Nucleophilic degradation of chloroalkenes by $O_2^{-\cdot}$*

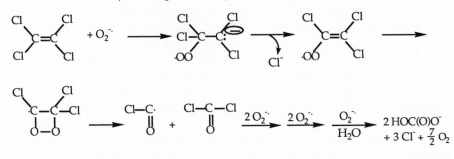

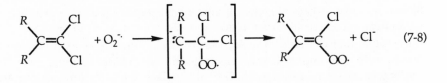

In the case of $(p\text{-ClPh})_2C=CCl_2$ (DDE) and 1,1-dichloroethene, addition of superoxide can only be followed by $\beta$-elimination of chloride if attack occurs at the carbon with the chlorine atoms [Eq. (7-8) and Scheme 7-6].

$$\text{R}\Big\rangle\!C\!=\!C\!\Big\langle^{\text{Cl}}_{\text{Cl}} + O_2^{-\cdot} \longrightarrow \left[ \begin{array}{c} \text{R} \quad \text{Cl} \\ \text{:}C\!-\!C\!-\!Cl \\ \text{R} \quad \text{OO}\cdot \end{array} \right] \longrightarrow \text{R}\Big\rangle\!C\!=\!C\!\Big\langle^{\text{Cl}}_{\text{OO}\cdot} + Cl^- \qquad (7\text{-}8)$$

Further reaction of the vinyl peroxy radical would be analogous to that shown in Scheme 7-7. This sequence would lead to the ketones [$RC(O)R$] that are observed as products (Table 7-1).

The apparent nonreactivity of *trans*-1,2-dichloroethene (and of Dieldrin and Aldrin) is not immediately explained by this mechanism in view of the facile reactivity of *cis*-1,2-dichloroethene. Steric effects may account for the differential reactivities of *cis*- and *trans*-1,2-dichloroethene; addition of superoxide ion would be expected to relieve steric strain between the two chlorines in the *cis* isomer. However, this reactivity pattern also is consistent with base-catalyzed elimination reactions of these systems.[32]

Hence, the elimination reaction (dehydrohalogenation) of *cis*-1,2-dichloroethene [Eq. (7-9)] is much more facile than that of the *trans* isomer because *trans* elimination can occur with the former but not the latter.[32] Subsequently, chloroethyne can react with superoxide ion via nucleophilic attack to give the observed products (Table 7-1).

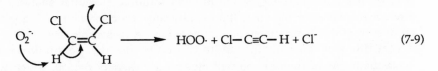

$$\qquad \longrightarrow \quad HOO\cdot + Cl\!-\!C\!\equiv\!C\!-\!H + Cl^- \qquad (7\text{-}9)$$

Both the nucleophilicity and basicity of $O_2^{-\cdot}$ are leveled by $Me_2SO$ (relative to DMF and acetonitrile),[33] which accounts for the slower reaction rates that are observed.

These observations indicate that in vivo superoxide ion could react with ingested chloroethenes. The proposed radical intermediates (vide supra) are likely toxins, and their reactivity with lipids may represent the mechanism for the cytotoxicity of chlorinated cleaning solvents in the liver.

Polyhalogenated aromatic hydrocarbons [e.g., hexachlorobenzene (HCB, $C_6Cl_6$) and polychorobiphenyls (PCBs)] are rapidly degraded by superoxide ion in dimethylformamide to bicarbonate and halide ions (Table 7-1).[23] Because halogen-bearing intermediates are not detected, the initial nucleophilic attack is the rate-determining step. The rates of reaction exhibit a direct correlation with the electrophilicity of the substrate (reduction potential) (e.g., $C_6Cl_6$, $E^{o'} = -1.48$ V versus SCE; $k_1/[S] = 1 \times 10^3\ M^{-1}s^{-1}$ and $1,2,4\text{-}C_6H_3Cl_3$, $E^{o'} = -2.16$ V; $k_1/[S] = 2 \times 10^{-2}$ $M^{-1}s^{-1}$).[23]

Although polyhaloaromatics are degraded by $O_2^{-\cdot}$ in acetonitrile and dimethyl sulfoxide, the rates of reaction are about one-tenth as great in MeCN and 20 times slower in $Me_2SO$. A reasonable initial step for these oxygenations is nucleophilic addition of $O_2^{-\cdot}$ to the polyhalobenzene (e.g., $C_6Cl_6$; Scheme 7-8). Subsequent loss of chloride ion will give a benzoperoxy radical, which will close on an adjacent aromatic carbon center and add a second $O_2^{-\cdot}$ to become a peroxo nucleophile that can attack the adjacent carbochlorine center with displacement of chloride ion and a highly electrophilic tetrachloro center. The latter undergoes facile reactions with $O_2^{-\cdot}$ to displace the remaining chloro atoms. Thus, Scheme 7-8 outlines a possible mechanism, but the fragmentation steps are speculative and not supported by the detection of any intermediate species.[34,35]

When Arochlor 1268 (a commercial PCB fraction that contains a mixture of $Cl_7$, $Cl_8$, $Cl_9$, and $Cl_{10}$ polychlorobiphenyls) is combined with excess $O_2^{-\cdot}$, the entire mixture is degraded. Samples taken during the course of the reaction confirm that (1) most heavily chlorinated members react first (the initial nucleophilic addition is the rate–determining step) and (2) all components are completely dehalogenated. Tests with other PCB mixtures establish that those components with three or more chlorine atoms per phenyl ring are completed degraded by $O_2^{-\cdot}$ within several hours.

The reaction stoichiometries, product profiles, and apparent second-order rate constants for the combination of perfluoroaromatic molecules (and several hydro and dihydro derivatives) with excess superoxide ion in dimethyl-formamide are summarized in Table 7-1. The primary product from the combination of $C_6F_6$ with 2 equivalents of $O_2^{-\cdot}$ is $C_6F_5OO^-$ on the basis of the F-NMR spectrum of the product solution and the mass spectrum for the major peak from the capillary GC of the product solution.[24] Similar analyses of the product solutions for the other fluoro substrates are consistent with a peroxide product from the displacement of a fluoride ion. A reasonable first step for these oxygenations is nucleophilic addition of $O_2^{-\cdot}$ to the polyfluoroaromatic. Subsequent loss of fluoride ion will give an aryl peroxy radical, which will be reduced by a second $O_2^{-\cdot}$ to the aryl peroxide product. This reaction sequence (with the initial nucleophilic displacement the rate-determining step) is analogous to that observed for chlorohydrocarbons and polychlorobenzenes (Scheme 7-8). However, the peroxo product of the latter systems is an effective nucleophile that attacks a second substrate molecule (or an adjacent aryl chlorine

*Scheme 7-8   Nucleophilic degradation of hexachlorobenzene (HCB) by $O_2^{-\cdot}$*

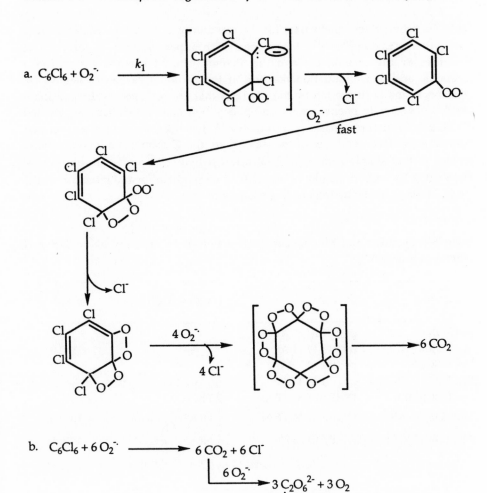

center). The stability of the $C_6F_5OO^-$ product indicates that it is unable to displace a fluoride from $C_6F_6$ or via an intramolecular process. As with other halocarbons, the apparent rate constants for the initial step of the $O_2^{-\cdot}/$ perfluoroaromatic reactions correlate with the reduction potentials for the substrates; the less negative the potential, the greater the reactivity. Nucleophilic attack of $C_6F_6$ by $HO^-$ to give $C_6F_5OH$ and $F^-$ is a well-documented process.[36]

The facile reactivity of perfluoroaromatic molecules with $O_2^{-\cdot}$ to give peroxy radicals and peroxides may represent a mode of cytotoxicity for such materials that parallels that for halogenated hydrocarbons, $C_6Cl_6$, and polychlorobiphenyls (PCBs).

*Addition to carbonyl centers*

Table 7-2 summarizes kinetic data for the reaction of $O_2^{-\cdot}$ with esters, diketones, and carbon dioxide.[35,37-39] Esters react with superoxide ion to form diacyl peroxides or the carboxylate and the alcohol. Initial reaction occurs via a reversible addition–elimination reaction at the carbonyl carbon (Scheme 7-9). This conclusion is supported by the products that are observed in the gas-phase reaction of $O_2^{-\cdot}$ with phenyl acetate and phenyl benzoate, which has been studied by Fourier-transform mass spectrometry.[40] In effect, there is a competition between loss of $O_2^{-\cdot}$ and loss of the leaving group. Carbanions are poor leaving groups, so that simple ketones without acidic $\alpha$-hydrogen atoms are unreactive. The $RC(O)OO\cdot$ radical should be a reactive intermediate for the initiation of the autoxidation of allylic hydrogens (see Chapter 5).

Table 7-2  Products and Kinetics for the Reaction of 1–5 m$M$ $O_2^{-\cdot}$ with Carbonyl Compounds at 25°C

| Substrate $S$ | Solvent$^a$ | [Products]/[S] | $k_2$, (M$^{-1}$ s$^{-1}$) |
|---|---|---|---|
| MeC(O)OEt | py/0.1 $M$ TEAP | — | 0.01 |
| MeC(O)OPh | py/0.1 $M$ TEAP | — | 160.0 |
| PhC(O)OPh | py/0.1 $M$ TEAP | — | 5.0 |
| PhCH(O) | py/0.1 $M$ TEAP | no reaction | — |
| PhC(O)C(O)Ph | DMF/0.1 $M$ TEAP | 2 PhC(O)O$^-$ | 2000.0$^b$ |
| MeC(O)C(O)Me | DMF/0.1 $M$ TEAP | $\frac{1}{2}$ HOOH, enolate | 4000.0$^b$ |
| MeC(O)C(O)OEt | DMF/0.1 $M$ TEAP | $\frac{1}{2}$ HOOH, enolate | 4000.0$^b$ |
| $CO_2$ | Me$_2$SO/0.1 $M$ TEAP | $\frac{1}{2}$ $^-$OC(O)OC(O)OO$^-$ | 1400.0$^b$ |

$^a$ py, pyridine; DMF, dimethylformamide; TEAP, tetraethylammonium perchlorate.
$^b$ Pseudo-first-order rate constants divided by substrate concentration, $k_1/[S]$, determined from measurements with a glassy-carbon–glassy-carbon ring-disk electrode that was rotated at 900 rev min$^{-1}$.

*Scheme 7-9  Nucleophilic hydrolysis and oxygenation of esters by $O_2^{-\cdot}$*

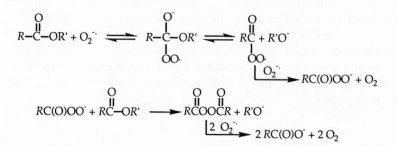

Simple diketones such as 2,3-butanedione are rapidly deprotonated by $O_2^{-\cdot}$, but the original diketone is recovered upon acidification (Scheme 7-10).

*Scheme 7-10   Enolization of 2,3-butanedione by $O_2^{-\cdot}$*

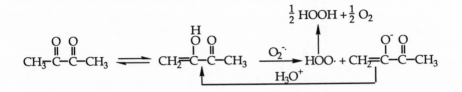

However, benzil [PhC(O)C(O)Ph] cannot enolize and is dioxygenated by $O_2^{-\cdot}$ to give two benzoate ions. Scheme 7-11 outlines a proposed mechanism that is initiated by nucleophilic attack. An alternative pathway has been proposed[41] in which the initial step is electron transfer from $O_2^{-\cdot}$ to the carbonyl, followed by coupling of the benzil radical with dioxygen to give the cyclic dioxetanelike intermediate.   However, the outer-sphere reduction potential for electron transfer from $O_2^{-\cdot}$ to a carbonyl carbon is insufficient (-0.60 V versus NHE, Chapter 2).

*Scheme 7-11   Nucleophilic dioxygenation of benzil by $O_2^{-\cdot}$*

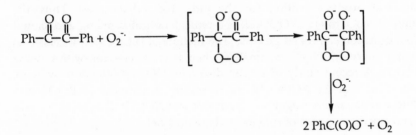

Carbon dioxide reacts rapidly with $O_2^{-\cdot}$ in aprotic solvents with a net stoichiometry in acetonitrile that is given by

$$2\,CO_2 + 2\,O_2^{-\cdot} \rightarrow C_2O_6^{2-} + O_2 \qquad (7\text{-}10)$$

and the proposed reaction path is outlined in Scheme 7-12. This reaction is significant because it provides a route to an activated form of carbon dioxide that may be involved in the vitamin K–dependent carboxylation of glutamic acid residues.[42] The results indicate that likely candidates for active intermediates are the anion radical, $CO_4^{-\cdot}$, or a hydrolysis product of $C_2O_6^{2-}$ such as peroxybicarbonate, $HOC(O)OO^-$.

*Scheme 7-12   Nucleophilic peroxidation of carbon dioxide by $O_2^{-\cdot}$*

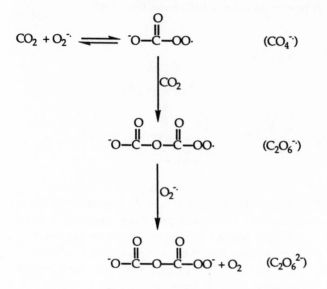

The primary product from the nucleophilic addition of $O_2^{-\cdot}$ to electrophilic substrates with leaving groups (*RX*) is a peroxy radical (*ROO·*). The presence of a large excess of substrate often precludes the diffusion-controlled reduction of *ROO·* by $O_2^{-\cdot}$, and results in the reaction manifold that is outlined in Scheme 7-13. The limits on the rate constants are consistent with the observed reactivities and product profiles for the indicated substrates.[43] Thus, the dimerization of *ROO·* to give *ROOOOR* is a radical–radical coupling and should be close to a diffusion-controlled process. The lifetime of [*ROOOOR*] is related to its O–O bond energies, which are smaller when *R* is an electron-withdrawing group (Cl₃C). A recent study of singlet dioxygen ($^1O_2$) production by these systems confirms that the *ROOOOR* intermediates decompose to *ROOR* and $^1O_2$,[31] while a separate investigation[43] indicates that *ROOOOR* is more reactive than $^1O_2$ and is a dioxygenase of rubrene and *cis*-stilbene.

### One-electron reductant

Another characteristic of $O_2^{-\cdot}$ is its ability to act as a moderate one-electron reducing agent. For example, combination of $O_2^{-\cdot}$ with 3,5-di-*tert*-butylquinone (DTBQ) in DMF yields the semiquinone anion radical DTBSQ$^{-\cdot}$ as the major product.[44] The relevant redox potentials in DMF are $O_2/O_2^{-\cdot}$, $E^{o\prime} = -0.60$ V versus NHE and DTBQ/DTBSQ$^{-\cdot}$, $E^{o\prime} = -0.25$ V versus NHE, which indicate that the equilibrium constant $K$ for the reaction of $O_2^{-\cdot}$ with DTBQ has a value of 0.8 x $10^6$.

$$O_2^{-\cdot} + DTBQ \rightleftharpoons DTBSQ^{-\cdot} + O_2 \qquad (7\text{-}11)$$

*Scheme 7-13 $(O_2^{\cdot -})$-induced formation of ROO· and its reaction paths*

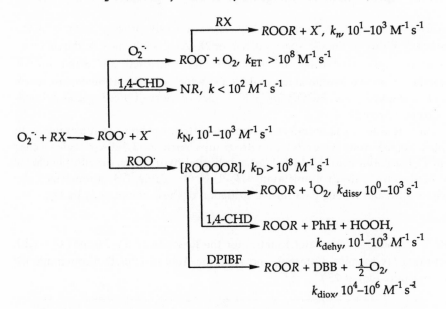

($RX = CCl_4$, $F_3CCCl_3$, $PhCCl_3$, $PhC(O)Cl$, BuBr, BuCl; 1,4-CHD=1,4-cyclohexadiene; DPIBF = diphenylisobenzofuran; DBB = dibenzoylbenzene.)

Electrochemical studies of sulfur dioxide[45] ($SO_2/SO_2^{\cdot -}$, $E^{\circ\prime} = -0.58$ V versus NHE) and of molecular oxygen[46] in dimethylformamide indicate that the equilibrium constant $K$ for the reaction of $SO_2$ with $O_2^{\cdot -}$ has a value of 1.1.

$$SO_2 + O_2^{\cdot -} \rightleftharpoons O_2 + SO_2^{\cdot -} \tag{7-12}$$

However, the reaction goes to completion because $SO_2^{\cdot -}$ is complexed by $SO_2$ and also dimerizes to dithionite ion.[47] Thus, $O_2^{\cdot -}$ is a stronger and more effective reducing agent than dithionite ion in aprotic solvents.

In aqueous media the equilibrium constant for the reduction of [iron(III)]-cytochrome *c* by $O_2^{\cdot -}$ has a value of $3.7 \times 10^4$.[48] Within DMF the reduction of [iron(III)]- porphyrin is even more complete.

$$(TPP)Fe^{III}(DMF)^+ + O_2^{\cdot -} \longrightarrow O_2 + (TPP)Fe^{II}, \; K, 7 \times 10^{11} \tag{7-13}$$

Superoxide ion is an effective reducing agent of transition-metal complexes; examples include copper(II),[49,50] manganese(III),[51,52] and iron(III).[53,54] It also reduces ferricenium ion, $Mn^{IV}_2O_2(o\text{-phen})_4^{4+}$, $Co^{III}(o\text{-phen})_3^{3+}$, and $Ir^{IV}Cl_6^{2-}$ by one-electron processes.[55]

## Sequential deprotonation–dehydrogenation of dihydrogroups.

The direct transfer of an electron to $O_2^-\cdot$ is an unlikely process because of the extreme instability of the $O_2^{2-}$ species (Chapter 2). As a result, most of the reported electron-transfer oxidations by $O_2^-\cdot$ actually represent an initial proton abstraction to give substrate anion and $HOO\cdot$, which disproportionates (or reacts with $O_2^-\cdot$) to give $O_2$ and $HOOH$ (or $HOO^-$). One or more of these species oxidizes the substrate anion.

Superoxide ion is an effective hydrogen-atom oxidant for substrates with coupled hetero-atom (O or N) dihydrogroups such as 3,5-di-*t*-butylcatechol ($DTBCH_2$), ascorbic acid ($H_2Asc$), 1,2-disubstituted hydrazines, dihydrophenazine ($H_2Phen$), and dihydrolumiflavin ($H_2Fl$).[56–58] Table 7-3 summarizes the stoichiometric and kinetic data for the oxidation of these compounds by $O_2^-\cdot$.

Table 7-3  Stoichiometries and Kinetics for the Reaction of 0.1–5.0 mM $O_2^-\cdot$ with Substituted Hydrazines, Catechol, and Ascorbic Acid in Dimethylformamide (or Dimethylsulfoxide) at 25°C

| Substrate $S$, 1–10 mM[a] | $[O_2^-\cdot]/[S]$ | [Anion radical]/$[S]$[b] | [HOOH]/$[S]$ | $k_1/[S]$[c] $(M^{-1}\,s^{-1})$ |
|---|---|---|---|---|
| PhNHNHPh | 1.0 | 1.0 | 1.0 | >100 |
| MeNHNHMe | 1.0 | 1.0 | 1.0 | > 30 |
| H$_2$Phen | 1.0 | 0.9 | 1.0 | >560 |
| H$_2$Fl | 1.0 | 0.8 | 0.9 | >340 |
| PhNHNH$_2$ | 2.0 | 0.0 | 2.0 | 110 |
| MeNHNH$_2$ | 2.0 | 0.0 | 2.0 | 16 |
| Ph$_2$NNH$_2$ | NR | | | |
| Me$_2$NNH$_2$ | NR | | | |
| DTBCH$_2$ | 1.0 | 0.8 | 0.9 | $1 \times 10^{4d}$ |
| H$_2$Asc | 1.0 | 0.8 | 0.9 | $2 \times 10^{4e}$ |

[a] $DTBCH_2$, 3,5-di-*tert*-butylcatechol; $H_2Asc$, ascorbic acid; $H_2Phen$, dihydrophenazine; $H_2Fl$, dihydrolumiflavin.

[b] The uv-visible absorption spectra for the anion radical products were compared with those for the products from controlled potential electrolytic reduction of 3,5-di-*tert*-butyl-*o*-benzoquinone ($\lambda_{max}$, 340 and 380 nm), dehydroascorbic acid ($\lambda_{max}$, 360 nm), phenazine, lumiflavin ($\lambda_{max}$, 420 nm), and azobenzene ($\lambda_{max}$, 410 nm).

[c] Pseudo-first-order rate constants, $k_1$ (normalized to unit substrate concentration $[S]$), were determined from current-collection-efficiency data ($O_2^-\cdot$ decay rates) with a glassy-carbon–glassy-carbon ring-disk electrode that was rotated at 900 rpm. ($O_2^-\cdot$ was produced at the disk electrode from dissolved $O_2$, which reacted with 2 mM substrate, and the unreacted $O_2^-\cdot$ was determined by its oxidation at the ring electrode.)

[d] $k_{bi} = 1 \times 10^4\,M^{-1}\,s^{-1}$ by stopped-flow spectrophotometry.

[e] $k_{bi} = 2.8 \times 10^4\,M^{-1}\,s^{-1}$ by stopped-flow spectrophotometry.

The general mechanism involves the rapid sequential transfer to $O_2^-$ of a proton and a hydrogen atom to form HOOH and the anion radical of the dehydrogenated substrate (Scheme 7-14).

*Scheme 7-14   Deprotonation–dehydrogenation of dihydro substrates by $O_2^-$*

*General initiation step:*

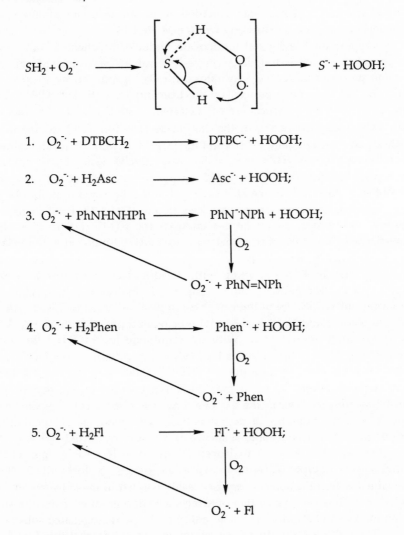

1.  $O_2^- + DTBCH_2 \longrightarrow DTBC^- + HOOH$;

2.  $O_2^- + H_2Asc \longrightarrow Asc^- + HOOH$;

3.  $O_2^- + PhNHNHPh \longrightarrow PhN^-NPh + HOOH$;

                              $\downarrow O_2$

                    $O_2^- + PhN=NPh$

4.  $O_2^- + H_2Phen \longrightarrow Phen^- + HOOH$;

                              $\downarrow O_2$

                    $O_2^- + Phen$

5.  $O_2^- + H_2Fl \longrightarrow Fl^- + HOOH$;

                              $\downarrow O_2$

                    $O_2^- + Fl$

With 1,2-diphenylhydrazine (PhNHNHPh) the azobenzene anion radi is rapidly oxidized by dioxygen to azobenzene

$$PhN^-NPh + O_2 \longrightarrow PhN=NPh + O_2^- \qquad (7\text{-}14)$$

which also is true for the phenazine and lumiflavin anion radicals.  Hence, $O_2^{-\cdot}$ acts as an initiator for the auto-oxidation of these compounds (see Scheme 7-14). For 1,2-diphenylhydrazine, turnover numbers in excess of 200 substrate molecules per $O_2^{-\cdot}$ have been observed.  The 1,2-diphenylhydrazine auto-oxidation cycle can be initiated by $HO^-$, which indicates that $O_2^{-\cdot}$ is formed in the ($HO^-$)-initiated process. Superoxide ion  also  initiates  the  auto-oxidation of dihydrophenazine, which is a model for dihydroflavin.  For example, the addition of 1 m$M$ $(Me_4N)O_2$ in DMF to 10 m$M$ $H_2$Phen in an $O_2$-saturated DMF solution results in the complete oxidation of the substrate (about 80% recovered as phenazine) and the production of 9–10 m$M$ HOOH.

Support for the general mechanism outlined in Scheme 7-14 is provided by gas–phase Fourier-transform mass-spectrometric studies of the anionic reaction products of several substrates with $O_2^{-\cdot}$ (produced by electron impact with $O_2$; $HO^-$ can be produced by electron impact with $H_2O$).[40,51] In these experiments neutral products are not detected.  Both $O_2^{-\cdot}$ and $HO^-$ react rapidly with 1,2-diphenylhydrazine in the gas phase ($P \approx 10^{-7}$ Torr) to give the anion radical of azobenzene (PhN$^{-\cdot}$NPh; $m/e^-$ = 182) and the anion from deprotonation (PhN$^-$NHPh; $m/e^-$ = 183), respectively. When $O_2^{-\cdot}$ is ejected from the experiment, the peak at $m/e^-$ = 182 disappears.  In contrast to the exponential decay that is observed for the $HO^-$ peak with time, the ion current for $O_2^{-\cdot}$ decays to a steady-state concentration.  Apparently, the PhN$^{-\cdot}$NPh product reacts with residual $O_2$ (which cannot be ejected from the FTMS cell) to give $O_2^{-\cdot}$ and azobenzene in a process that is analogous to the ($O_2^{-\cdot}$)-induced auto-oxidation in aprotic solvents (Scheme 7-14).

Analogous FTMS studies with 1,2-dihydroxybenzenes also provide support for the general mechanism. Superoxide ion reacts rapidly with 3,5-di-*t*-butylcatechol (DTBCH$_2$) in the gas phase to give the anion (DTBCH$^-$; $m/e^-$ = 221) and the anion radical of 3,5-di-*t*-butyl-*o*-benzoquinone (DTBSQ$^{-\cdot}$; $m/e^-$ = 220) in an approximate ratio of 3:1.  With hydroquinone [*p*-Ph(OH)$_2$] the dominant product (ca. 70%) is the anion radical (SQ$^{-\cdot}$; $m/e^-$ = 108).  When $HO^-$ is the gas-phase reagent, the only product for DTBCH$_2$ [and for *o*-Ph(OH)$_2$] is the anion from deprotonation.  The fact that the anion radicals of the dehydrogenated substrates are produced in the gas phase as well as in aprotic solvents confirms that the reaction sequence of deprotonation/hydrogen-atom abstraction to form HOOH must either be a rapid-sequence or a nearly concerted process.  Because $O_2^{-\cdot}$ is expected to abstract hydrogen atoms much less easily than HOO·, the initial step is deprotonation of the substrate by $O_2^{-\cdot}$ to form HOO·; the latter (contained within the solvent cage or weakly bonded to the substrate in a "sticky collision" in the gas phase) then abstracts a hydrogen atom from the substrate anion to form HOOH and the anion radical of the dehydrogenated substrate.

Thus, the ($O_2^{-\cdot}$)-induced autoxidations of 1,2-disubstituted hydrazines, dihydrophenazines, and dihydroflavins in aprotic media provide a simple pathway for rapid conversion of dioxygen to HOOH, and one that does not involve catalysis by metal ions or metalloproteins. This is exemplified by the net reaction for dihydrolumiflavin,

$$H_2Fl + O_2 \xrightarrow{\;O_2^{-\cdot},\,HO^-\;} Fl + HOOH \tag{7-15}$$

The overall reaction stoichiometry and products from the combination of $O_2^{-\cdot}$ and phenylhydrazine (PhNHNH$_2$) in Me$_2$SO are represented by (Table 7-3)

$$PhNHNH_2 + 2\,O_2^{-\cdot} \longrightarrow PhH + N_2 + 2\,HOO^- \tag{7-16}$$

Because half of the PhNHNH$_2$ remains unreacted from a 1:1 $O_2^{-\cdot}$/PhNHNH$_2$ combination, the primary step must be rate limiting and followed by a rapid step that consumes a second $O_2^{-\cdot}$. Second-order kinetics are observed by the rotated ring-disk voltammetric experiment,[19] and molecular oxygen is not detected during the course of the process or as a product. A reasonable reaction sequence that is consistent with the experimental results has H-atom abstraction by $O_2^{-\cdot}$ as the primary rate-limiting step

$$PhNHNH_2 + O_2^{-\cdot} \longrightarrow Ph\overset{\cdot}{N}NH_2 + HOO^- \tag{7-17}$$

followed by rapid coupling of the radical product with a second $O_2^{-\cdot}$ and degradation to the final products.

$$[PhN=NH] + HOO^-$$
$$PhH + N_2 \tag{7-18}$$

The limited yield of biphenyl (<5%), the trace amounts of trapped phenyl radical, and the absence of oxygenated phenyl products are consistent with the reaction sequence of Eqs. (7-17) and (7-18), and with the reaction intermediate of Eq. (7-18). The reaction of $O_2^{-\cdot}$ with methylhydrazine is analogous to that for phenylhydrazine with methane the major product. The slower rate is consistent with the enhanced stabilization of radical intermediates by aromatic substituents.

Unsymmetrical disubstituted hydrazines (1,1-diphenylhydrazine and 1,1-dimethylhydrazine) do not exhibit a detectable reaction with $O_2^{-\cdot}$ in the gas phase nor in Me$_2$SO within approximately 10 min. This is consistent with the hypothesis that only those substituted hydrazines with a secondary amine function react with $O_2^{-\cdot}$ by hydrogen-atom transfer. Although there is a report[59] that 1,1-diphenylhydrazine reacts with superoxide ion to produce N-nitrosodiphenylamine, such a product is more consistent with a peroxide reaction (a likely possibility for the reaction conditions).

The primary process for the oxidation by $O_2^{-\cdot}$ of dihydrophenazine and dihydro-lumiflavin must be analogous to that for PhNHNHPh (Scheme 7-14) to give the anion radicals of phenazine (Phen$^{-\cdot}$) and lumiflavin (Fl$^{-\cdot}$). These in turn react with $O_2$ to give $O_2^{-\cdot}$ plus phenazine and lumiflavin, respectively; the process is analogous to that for the anion radical of azobenzene (PhN$^{-\cdot}$NPh). The oxidation potentials ($E_{p,a}$) for PhN$^{-\cdot}$NPh, Phen$^{-\cdot}$ and Fl$^{-\cdot}$ in Me$_2$SO are -1.1 V versus NHE, -0.9 V, and -0.6 V, respectively.[19] Each value is sufficiently negative to reduce $O_2$ to $O_2^{-\cdot}$ (-0.5 V versus NHE in Me$_2$SO). Hence, the ($O_2^{-\cdot}$)-induced auto-oxidation of PhNHNHPh also is thermodynamically feasible for dihydrophenazine and dihydro-lumiflavin and does occur for these two model substrates of reduced flavoproteins (Table 7-3). Such an auto-oxidation reaction sequence may be relevant to the fractional yield of $O_2^{-\cdot}$ from the flavin-mediated activation of $O_2$[60] and the auto-oxidation of xanthine [catalyzed by xanthine oxidase (XO), a flavoprotein].[61]

$$Fl + H_2Sub \xrightarrow{\text{XO}} H_2Fl + Sub \qquad (7\text{-}19)$$

$$H_2Fl + Fl \underset{\xrightarrow{\hspace{1cm}}}{\overset{\text{HO}^-}{\rightleftharpoons}} 2\,Fl^{-\cdot} + 2\,H_2O \qquad (7\text{-}20)$$

$$Fl^{-\cdot} + O_2 \longrightarrow Fl + O_2^{-\cdot} \qquad (7\text{-}21)$$

$$H_2Fl + O_2^{-\cdot} \longrightarrow Fl^{-\cdot} + HOO \qquad (7\text{-}22)$$

The reaction of N-methylhydrophenazine with $O_2^{-\cdot}$ (in the absence of oxygen) yields N-methylphenazine radical (MePhen$\cdot$).

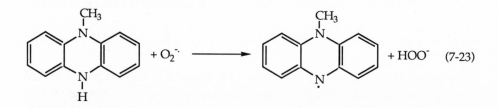

$$(7\text{-}23)$$

Perhaps the most significant aspect of the superoxide-, hydroxide-, or electron-induced auto-oxidation of donor molecules (1,2-diphenylhydrazine, dihydrophenazine, dihydro-lumiflavin, and reduced flavoproteins) is the activation of dioxygen to hydrogen peroxide in biological matrices. Thus, within the normal cytochrome $P$-450 metabolic cycle, either hydroxide ion or an electron-transfer co-factor acts as an initiator (probably to produce Fl$^{-\cdot}$) and reduced flavoproteins act as the H-atom donor.[60] In contrast, the introduction of $O_2^{-\cdot}$ or hydrated electrons (from ionizing radiation or a disease state) into a biological matrix that contains donor molecules leads to the uncontrolled formation of hydrogen peroxide. If reduced metal ions are present, Fenton chemistry occurs to give hydroxyl radicals, which will initiate lipid peroxidation[62] and rancidification of stored foodstuffs.

On the basis of their redox thermodynamics and reaction chemistry with $O_2^{-\cdot}$ in aprotic media, ascorbic acid[58] and some catechols[46] may be subject to an $(O_2^{-\cdot})$-catalyzed auto-oxidation to dehydroascorbic acid and *o*-quinones, respectively.

## Radical-radical coupling

Combination of superoxide ion with the cation radical $(MV^{+\cdot})$ of methyl viologen (1,1'-dimethyl-4,4'-bipyridinium ion, Paraquat) is one of the best-documented examples of a stoichiometric $O_2^{-\cdot}$ radical coupling reaction.[63,64] The initial addition to give an unstable diamagnetic product is assumed to be at $\alpha$-carbon atoms (which possess the maximum unpaired spin density in the cation radical), followed by formation of a dioxetanelike intermediate that decomposes to a complex mixture of products (Scheme 7-15). This process may provide a means to explain the mechanism of Paraquat toxicity. Several other reports have proposed direct coupling of $O_2^{-\cdot}$ to cation radicals.[65]

*Scheme 7-15  Oxygenation of methyl viologen cation radical via radical–radical coupling with $O_2^{-\cdot}$*

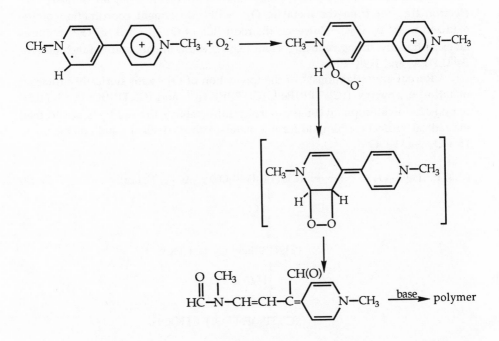

The $N^5$-ethyl-3-methyllumiflavo radical couples with $O_2^{-\cdot}$ to form the $N^5$-ethyl-4a-hydroperoxy-3-methyllumiflavin anion.[66]

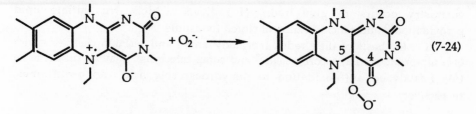

$$\text{(7-24)}$$

Because the 4a-flavin hydroperoxides are intermediates in both flavoprotein oxidase and mono-oxygenase chemistry,[67] possible mechanisms for their formation are fundamental to an understanding of their biogenesis.

Although superoxide ion reacts with reduced transition-metal complexes [Fe(II), Mn(II), Co(II), $Fe^{II}(EDTA)$, $Fe^{II}(TPP)$][55,68] to give oxidized products, for simple solvated cations the reaction sequence is a Lewis-acid-catalyzed disproportionation.[55]

$$Fe^{2+} + 2\,O_2^{-\cdot} + 2\,H_2O \longrightarrow O_2 + HOOH + Fe^{II}(OH)_2 \qquad \text{(7-25)}$$

with subsequent Fenton chemistry

$$2\,Fe^{II}(OH)_2 + HOOH \longrightarrow 2\,Fe^{III}(OH)_3(s) \qquad \text{(7-26)}$$

In the cases of $Fe^{II}(EDTA)$ and $Fe^{II}(TPP)$, the peroxide products $[Fe^{III}(EDTA)\text{-}(O_2^-)$[68] and $Fe^{III}(TPP)(O_2^-)$[54]$]$ were believed to result from an outer-sphere electron transfer from the metal to $O_2^{-\cdot}$ with subsequent coordination of the oxidized metal by $O_2^{2-}$. However, the reduction of $O_2^{-\cdot}$ to "naked" $O_2^{2-}$ requires an exceptionally strong reducing agent (e.g., sodium metal), and is precluded for $Fe^{II}(EDTA)$ and $Fe^{II}(TPP)$.

Recent investigations[69] of the interaction of $O_2^{-\cdot}$ with sterically protected metallo-porphyrins $[(Cl_8TPP)Fe^{II}, (Cl_8TPP)Mn^{II},$ and $(Cl_8TPP)Co^{II}; Cl_8TPP=$ tetrakis(2,6-dichlorophenyl)-meso-porphyrin] establish that adducts are formed via radical–radical coupling to form a metal–oxygen covalent bond (analogous to $H\cdot + O_2^{-\cdot} \rightarrow H\text{-}OO^-$)

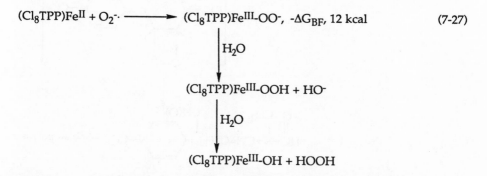

$$\text{(7-27)}$$

Similar reactivities occur with the manganese and cobalt porphyrins to give $(Cl_8TPP)Mn^{III}\text{-}OO^-$ ($-\Delta G_{BF}$, 17 kcal) and $(Cl_8TPP)Co^{III}\text{-}OO^-$ ($-\Delta G_{BF}$, 8 kcal). Such radical–radical coupling is favored for all reduced transition-metal complexes with unpaired electrons and probably is the reaction path for $O_2^{-\cdot}$ with $Fe^{II}(EDTA)$[68] and $Fe^{II}(TPP)$.[54]

Thus, the radical–radical coupling reaction path is favored under neutral conditions with reduced transition-metal complexes, and provides insight to the mechanisms for superoxide-ion disproportionation that are catalyzed by the iron and manganese superoxide dismutase proteins. The combination of $(Cl_8TPP)Fe^{II}$ or $(Cl_8TPP)Mn^{II}$ with $O_2^{-\cdot}$ does not result in electron transfer from the metal; instead they couple to form $(Cl_8TPP)Fe^{III}\text{-}OO^-$ and $(Cl_8TPP)Mn^{III}\text{-}OO^-$. The latter abstract protons from the medium, and the $(Cl_8TPP)FeOOH$ and $(Cl_8TPP)MnOOH$ products react with a second $O_2^{-\cdot}$ and a proton to give $HOOH$ and $O_2$

$$(Cl_8TPP)Fe^{II} + O_2^{-\cdot} + HA \longrightarrow (Cl_8TPP)Fe^{III}\text{-}OOH + A^- \qquad (7\text{-}28)$$

$$\Big\downarrow O_2^{-\cdot},\, HA$$

$$(Cl_8TPP)Fe^{II} + O_2 + HOOH + A^-$$

Hence, $(Cl_8TPP)Fe^{II}$ and $(Cl_8TPP)Mn^{II}$ facilitate the disproportionation of $O_2^{-\cdot}$, which is equivalent to the function of the iron and manganese superoxide dismutase proteins. Whether the mechanism of Eq. (7-28) is relevant to those for the proteins is unknown, but the absence of electron transfer from their metal centers to $O_2^{-\cdot}$ is a reasonable expectation (as is radical–radical coupling of $O_2^{-\cdot}$ and the protein in the primary step of the disproportionation mechanism).

**Biological systems**

*Biogenesis of $O_2^{-\cdot}/HOO\cdot$*

Aerobic organisms produce minor fluxes of superoxide ion during respiration and oxidative metabolism. Thus, up to 15% of the $O_2$ reduced by cytochrome-*c* oxidase and by xanthine oxidase passes through the $HOO\cdot/O_2^{-\cdot}$ state.[70] The reductase of the latter system is a flavoprotein[71] that probably reduces $O_2$ to $HOOH$ via a redox cycle similar to that outlined by Eqs. (7-19) – (7-22).[61] Thus, the observed flux of $O_2^{-\cdot}$, which is the carrier of the auto-oxidation cycle, is due to leakage during turnover of xanthine/xanthine oxidase (see Scheme 7-14 for a reasonable mechanistic pathway).

Because $HOO\cdot$ in water is a moderately weak acid ($pK_{HA}$, 4.9), at pH 7 about 1% of the biogenerated $O_2^{-\cdot}$ exists as $HOO\cdot$. The latter species is an effective initiator for the auto-oxidation of lipids via their allylic carbons (see discussion in Chapter 5), and probably represents the primary hazard of superoxide production in biological matrices.

*Superoxide dismutase chemistry, mechanisms, and function*

Most aerobic organisms have one or more of the three metalloproteins (manganese-SOD in mitochondria and prokaryotes, iron-SOD in plants and prokaryotes, and copper/zinc-SOD in red blood cells) whose purported function is the disproportionation of superoxide (superoxide dismutases).[61,70] Biologists are prone to believe that superoxide ion ($O_2^{-\cdot}$) is an effective free radical because it contains an unpaired electron. However, the discussion in Chapter 3 establishes that $O_2^{-\cdot}$ is unable to break any C–H bond under ambient conditions [the weakest C–H bond is for the allylic carbons in linoleic acid; $\Delta H_{DBE}$, ~74 kcal; the radical character of $O_2^{-\cdot}$ is represented by the H–OO$^-$ bond energy (-$\Delta G_{BF}$, 72 kcal)]. In contrast, the conjugate acid of $O_2^{-\cdot}$ (HOO·) is able to initiate auto-oxidation of allylic carbons (see Chapter 5) [the radical character of HOO· is represented by the H–OOH bond energy (-$\Delta G_{BF}$, 82 kcal)].

A second belief of most biologists is that the superoxide dismutase proteins safely destroy $O_2^{-\cdot}$ via electron-transfer cycles at their transition-metal centers, for example,

$$L_n Mn^{II}Y + O_2^{-\cdot} \rightarrow L_n Mn^{II} + O_2 + Y^- \tag{7-29}$$

$$L_n Mn^{II} + O_2^{-\cdot} + HY \rightarrow L_n Mn^{III}Y + HOOH \tag{7-30}$$

However, several studies[52,69] of model complexes indicate that radical–radical coupling (trapping $O_2^{-\cdot}$/HOO·) is the most reasonable mechanism, particularly in view of the need for an extremely fast rate for the protein/$O_2^{-\cdot}$ reaction ($k_{SOD}$, ~3 x $10^9$ $M^{-1}$ s$^{-1}$). Thus, a proposed two-step cycle does not involve electron transfer at the metal

$$L_n Mn^{II} + O_2^{-\cdot}/HOO· \xrightarrow[A^-]{HA} L_n Mn^{II}\text{-OOH} \tag{7-31}$$

with subsequent branch:
$O_2^{-\cdot}$/HOO·
$HA$
$A^-, k, 3 \times 10^{10} M^{-1}s^{-1}$
$L_n Mn^{II} + HOOH + O_2$

The advantage of such a cycle is that it traps $O_2^{-\cdot}$/HOO· via bond formation at diffusion controlled rates for subsequent reaction with a second $O_2^{-\cdot}$/HOO·.

This mechanistic proposal in turn prompts the suggestion that the function of the superoxide dismutase proteins is to prevent free HOO· from coming into contact with allylic C–H bonds in the biological matrix. One approach is to minimize the lifetime of $O_2^{-\cdot}$/HOO·, which is in addition to the radical-radical coupling proposition to deactivate HOO·. For a steady-state flux of 30 x $10^{-6}$ $M$ $O_2^{-\cdot}$/HOO· at pH 5: (1) without superoxide dismutase (SOD) the approximate half-life of $O_2^{-\cdot}$/HOO· is about 30 ms

$$\left\{ 2\,O_2^{-} \xrightarrow[A^-]{HA} [HOO\cdot + O_2^{-}] \xrightarrow[A^-]{HA,\,k} HOOH + O_2,\ k = 10^8\,M^{-1}\,s^{-1} \right\},$$

with free HOO· an essential component of the rate-determining step for the disproportionation reaction; and (2) with SOD ($k$, $3 \times 10^9\,M^{-1}\,s^{-1}$) via the reaction sequence of Eq. (7-31) the approximate half-life of $O_2^{-}$/HOO· is about 10 µs, and free HOO· is not required in the rate-determining step.

## References

1. Roberts, J. L., Jr.; Morrison, M. M.; Sawyer, D. T. *J. Am. Chem. Soc.* **1978**, *100*, 329.
2. Sawyer, D. T. *J. Phys. Chem.* **1989**, *93*, 7977.
3. Bielski, B. H. J. *Photochem. Photobiol.* **1978**, *28*, 645.
4. Bielski, B. H. J.; Allen, A. O. *J. Phys. Chem.* **1977**, *81*, 1048.
5. Ilan, Y. A.; Meisel, D.; Czapski, G. *Isr. J. Chem.* **1974**, *12*, 891.
6. Behar, D.; Czapski, G.; Rabini, J.; Dorfman, L. M.; Schwartz, H. A. *J. Phys. Chem.* **1970**, *74*, 3209.
7. Rabini, J.; Nielsen, S. O. *J. Phys. Chem.* **1969**, *73*, 3736.
8. Sawyer, D. T.; Nanni, E. J., Jr. In *Oxygen and Oxy-Radicals in Chemistry and Biology* (Powers, E. J.; Rodgers, M. A. J., eds.). New York: Academic Press, 1981.
9. Roberts, J. L., Jr.; Sawyer, D. T. *Isr. J. Chem.* **1983**, *23*, 430.
10. Chin, D. -H.; Chiericato, G., Jr.; Nanni, E. J., Jr.; Sawyer, D. T. *J. Am. Chem. Soc.* **1982**, *104*, 1296.
11. Gibian, M. J.; Sawyer, D. T.; Ungermann, T.; Tangpoonpholvivat, R.; Morrison, M. M. *J. Am. Chem. Soc.* **1979**, *101*, 640.
12. Nanni, E. J., Jr.; Stallings, M. D.; Sawyer, D. T. *J. Am. Chem. Soc.* **1980**, *102*, 4481.
13. Dietz, R.; Forno, A. E. J.; Larcombe, B. E.; Peover, M. D. *J. Chem. Soc. B* **1970**, 816.
14. Merritt, M. V.; Sawyer, D. T. *J. Org. Chem.* **1970**, *35*, 2157.
15. Magno, F.; Seeber, R.; Valcher, S. *J. Electroanal., Chem.* **1977**, *83*, 131.
16. San Fillipo, J., Jr.; Chern, C.-I.; Valentine, J. S. *J. Org. Chem.* **1975**, *40*, 1678.
17. Johnson, R. A.; Nidy, E. G. *J. Org. Chem.* **1975**, *40*, 1680.
18. Gibian, M. J.; Tangpoonpholvivat, R.; Ungermann, T.; Morrison, M. M.; Sawyer, D. T. Unpublished data.
19. Roberts, J. L., Jr.; Calderwood, T. S.; Sawyer, D. T. *J. Am. Chem. Soc.* **1983**, *105*, 7691.
20. Calderwood, T. S.; Neuman, R. C., Jr.; Sawyer, D. T. *J. Am. Chem. Soc.* **1983**, *105*, 3337.
21. Calderwood T. S. Sawyer D. T. *J Am Chem. Soc.* **1984**, *106*, 7185.
22. Hojo, M. Sawyer D. T. *Chem. Res. Toxicol.* **1989**, *2*, 193.
23. Sugimoto, H.; Matsumoto, S.; Sawyer, D. T. *J. Am. Chem. Soc.* **1987**, *109*, 8081.
24. Sugimoto, H.; Matsumoto, S.; Sawyer, D. T.; Kanofsky, J. R. *J. Am. Chem. Soc.* **1988**, *110*, 5193.
25. Merritt, M. V.; Johnson, R. A. *J. Am. Chem. Soc.* **1977**, *99*, 3713.
26. Eberson, L. *Adv. Phys. Org. Chem.* **1982**, *18*, 79.

27. Mason, R. P. In *Free Radicals in Biology, Vol. 6* (Pryor, W. A., ed.). New York: Academic Press, 1982, pp. 161–222.

28. Slater, T. F. In *Free Radicals, Lipid Peroxidation and Cancer* (McBrien, D. C. H.; Slater, T. F., ed.). New York: Academic Press, 1982, pp. 243–274.

29. Perrin, C. L. *J. Phys. Chem.* **1984**, *88*, 3611.

30. Oae, S.; Takata, S. *Tetrahedron Lett.* **1980**, *21*, 3689.

31. Kochi, J. K. In *Free Radicals* (Kochi, J. K., ed.). New York: Wiley, 1973, p. 698.

32. Truce, W. E.; Boudakian, M. M.; Heine, R. F.; McManimie, R. J. *J. Am. Chem. Soc.* **1956**, *78*, 2743.

33. Sawyer, D. T.; Chiericato, G., Jr.; Angelis, C. T.; Nanni, E. J., Jr.; Tsuchiya, T. *Anal. Chem.* **1982**, *54*, 1720.

34. Sawyer, D. T.; Stamp, J. J.; Menton,K. A. *J. Org. Chem.* **1983**, *48*, 3733.

35. Roberts, J. L., Jr.; Sawyer, D. T. *J. Am. Chem. Soc.* **1984**, *106*, 4667.

36. Kununyants, I. L.; Yakobson, G. G., (eds.). *Syntheses of Fluoroorganic Compounds.* New York: Springer-Verlag, 1985.

37. Gibian, M. J.; Sawyer, D. T.; Ungermann, T.; Tangpoonpholvivat, R.; Morrison, M. M. *J. Am. Chem. Soc.* **1979**, *101*, 640.

38. Magno, F.; Bontempelli, G. *J. Electroanal. Chem.* **1976**, *68*, 337.

39. Sawyer, D. T.; Stamp, J. J.; Menton, K. A. *J. Org. Chem.* **1983**, *48*, 3733.

40. Johlman, C. L.; White R. L.; Sawyer, D. T.; Wilkins, C. L. *J. Am. Chem. Soc.* **1983**, *105*, 2091.

41. Frimer, A. A. In *The Chemistry of Functional Groups, Peroxides* (Patai, S., ed.). New York: Wiley, 1983; pp. 429–461.

42. Esnouf, M. P.; Green, M. R.; Hill, H. A. O., Irvine, G. B.; Walter, S. J. *Biochem. J.* **1978**, *174*, 345.

43. Matsumoto, S.; Sugimoto, H.; Sawyer, D. T. *Chem. Res. Toxicol.* **1988**, *1*, 19.

44. Sawyer, D. T.; Richens, D. T.; Nanni, E. J., Jr.; Stallings, M. D. *Dev.Biochem.* **1980**, *11A*, 1.

45. Martin, R. P.; Sawyer, D. T. *Inorg. Chem.* **1972**, *11*, 2644.

46. Nanni, E. J., Jr.; Stallings, M. D.; Sawyer, D. T. *J. Am. Chem. Soc.* **1980**, *102*, 4481.

47. Stallings, M. D.; Sawyer, D. T. *J. Chem. Soc., Chem. Commun.* **1979**, *340*.

48. Rao, P. S.; Hazen, E. *J. Phys. Chem.* **1975**, *79*, 397.

49. Valentine, J. S.; Curtis, A. B. *J. Am. Chem. Soc.* **1975**, *97*, 224.

50. Klug-Roth, D.; Rabini, J. *J. Phys. Chem.* **1976**, *80*, 588.

51. Valentine, J. S.; Quinn, A. E. *Inorg. Chem.* **1976**, *15*, 1977.

52. Howie, J. K.; Morrison, M. M., Sawyer, D. T. *ACS Symp. Ser.* **1977**, No. *38*, 97.

53. Halliwell, B. *FEBS Lett.* **1975**, *56*, 34.

54. McCandlish, E.; Miksztal, A. R.; Nappa, M.; Sprenger, A. Q.; Valentine, J. S.; Strong, J. D.; Spiro, T. G. *J. Am. Chem. Soc.* **1980**, *102*, 4268.

55. Nanni, E. J., Jr. Ph.D. Dissertation, University of California, Riverside, CA, 1980.

56. Sawyer, D. T.; Calderwood, T. S.; Johlman, C. L.; Wilkins, C. L. *J. Org. Chem.* **1985**, *50*, 1409.

57. Calderwood, T. S.; Johlman, C. L.; Roberts, J. L., Jr.; Wilkins, C. L.; Sawyer, D. T. *J. Am. Chem. Soc.* **1984**, *106*, 4683.

58. Sawyer, D. T.; Chiericato, G., Jr.; Tsuchiya, T. *J. Am. Chem. Soc.* **1982**, *104*, 6273.

59. Chern, C.-I.; San Filippo, J. *J. Org. Chem.* **1977**, *42*, 178.

60. Hemmerich, P.; Wessick, A. In *Biochemical and Clinical Aspects of Oxygen* (Caughey, W. S., ed.). New York: Academic Press, 1979; pp. 491–512.
61. McCord, J. M.; Fridovich, I. *J. Biol. Chem.* **1968**, *243*, 5753.
62. Aust, S. D.; Svingen, B. A. In *Free Radicals in Biology* (Pryor, W. A., ed.). New York: Academic Press, 1982, Vol. V, pp. 1–28.
63. Nanni, E. J., Jr.; Sawyer, D. T. *J. Am. Chem. Soc.* **1980**, *102*, 7591.
64. Nanni, E. J., Jr.; Angelis, C. T.; Dickson, J.; Sawyer, D. T. *J. Am. Chem. Soc.* **1981**, *103*, 4268.
65. Ando, W.; Kabe, Y.; Kobayashi, S., Takyu, C.; Yamagishi, A.; Inaba, H. *J. Am. Chem. Soc.* **1980**, *102*, 4527.
66. Nanni, E. J., Jr.; Sawyer, D. T.; Ball, S. S.; Bruice, T. C. *J. Am. Chem. Soc.* **1981**, *103*, 2797.
67. Walsh, C. *Acc. Chem. Res.* **1980**, *13*, 148.
68. McClure, G. J.; Fee, J. A.; McCluskey, G. A.; Groves, J. T. *J. Am. Chem. Soc.* **1977**, *99*, 5220.
69. Tsang, P. K. S.; Sawyer, D. T. *Inorg. Chem.* **1990**, *29*, 2848.
70. Fridovich, I. *J. Biol. Chem.* **1970**, *245*, 4053.
71. Fridovich, I. In *Advances in Organic Biochemistry* (Eichhorn, G. L.; Marzilli, D. L., eds.). New York: Elsevier–North Holland, 1979, pp. 67–90.

# 8

# REACTIVITY OF OXY-ANIONS
## [HO⁻ (RO⁻), HOO⁻ (ROO⁻), AND O₂⁻·] AS NUCLEOPHILES AND ONE-ELECTRON REDUCING AGENTS

*The electron is the ultimate Lewis base and nucleophile: [$e^- + H_2O \equiv HO^- + H\cdot$]*

The preceding chapter describes the primary reaction chemistry of superoxide ion ($O_2^{-\cdot}$) to be that of (1) a Brønsted base (proton transfer from substrate), (2) a nucleophile (via displacement or addition), (3) a one-electron reductant, and (4) a dehydrogenase of secondary-amine groups. The chemistry is characteristic of all oxy anions [HO⁻ (RO⁻), HOO⁻ (ROO⁻), and $O_2^{-\cdot}$], but the relative reactivity for each is determined by its $pK_a$ and one-electron oxidation potential, which are strongly affected by the anionic solvation energy of the solvent matrix. The present chapter will focus on the reactivity of hydroxide ion (HO⁻), but the principles apply to all oxy anions and permit assessments of their relative reactivity.

The reactivity of hydroxide ion (and that of other oxy anions) is interpreted in terms of two unifying principles: (1) the redox potential of the YO⁻/YO· (Y = H, R, HO, RO, and O) couple (in a specific reaction) is controlled by the solvation energy of the YO⁻ anion and the bond energy of the R-OY product ($RX + YO^- \rightarrow R\text{-}OY + X^-$), and (2) the nucleophilic displacement and addition reactions of YO⁻ occur via an inner-sphere *single-electron* shift.[1] The electron is the ultimate base and one-electron reductant, which, upon introduction into a solvent, is transiently solvated before it is "leveled" (reacts) to give the conjugate base (anion reductant) of the solvent. Thus, in water the hydrated electron ($e^-$)$_{H_2O}$ yields HO⁻ via addition to the H–OH bond of water[2]

$$(e^-)_{H_2O} + H\text{–}OH \rightarrow H\cdot + (HO^-)_{H_2O}, \quad E°, \text{-}2.93 \text{ V versus NHE} \quad (8\text{-}1)$$

The product combination (H· + HO⁻) represents the ultimate thermodynamic reductant for aqueous systems. In the absence of an H· atom (and the stabilization afforded to HO· by formation of the 119-kcal H–OH bond),[3] the hydroxide ion becomes a much less effective reductant[2]

$$(e^-)_{H_2O} + HO\cdot \rightarrow (HO^-)_{H_2O}, \quad E°, \text{+}1.89 \text{ V versus NHE} \quad (8\text{-}2)$$

### Solvent effects on the redox chemistry of HO⁻

Table 8-1 summarizes the redox potentials in water and in acetonitrile for the single-electron oxidation of HO⁻ and other bases.[2,4] In MeCN the HO·/HO⁻ redox potential is more negative by about 1.0 V and the $O_2/O_2^{-\cdot}$ redox potential by about 0.5 V relative to their values in $H_2O$. Most of this is due to the decrease in the energy of solvation for $HO^-$ and $O_2^{-\cdot}$ in MeCN (compared to water, where each has an estimated energy of hydration of about 100 kcal/mol).[5] The increase in the ionization energy for $HO^-$ from 1.8 eV in the gas phase to 6.2 eV in water[5] attests to its large solvation energy and to its dramatic deactivation as a base and nucleophile in water.[6-10]

Hydroxide ion is a stronger base and a better one-electron donor in MeCN and $Me_2SO$ than in water, because these organic solvents have solvation energies for $HO^-$ that are 20–25 kcal $mol^{-1}$ (~1 eV) less than water.[11] Thus, reduced solvation of $HO^-$ decreases its ionization energy and causes it to have a more negative redox potential and to be a stronger electron donor. (The solvation energies of the $HO\cdot$ radical and neutral molecules are small,[6] so that the large changes in redox potential are due primarily to the solvation of anions.) This dramatically enhances the reactivity of $HO^-$ toward the electrophiles shown in Table 8-1b and reveals a facet of $HO^-$ chemistry that is effectively quenched by water, namely its ability to function as a one-electron reducing agent.

The redox potentials for the electron acceptors that react with $HO^-$ (Table 8-1) are such that a pure outer-sphere single-electron transfer (SET) step would be endergonic (the $HO\cdot/HO^-$ redox potential is more positive than the redox potential of the electron acceptor). Hence, the observed net reactions must be driven by coupled chemical reactions; particularly bond formation by the $HO\cdot$ to the electrophilic atom of the acceptor molecule that accompanies a single-electron shift. (The formation of the bond provides a driving force sufficient to make the overall reaction thermoneutral or exergonic; 1.0 V per 23.1 kcal of bond energy.) A recent study of the effect of various transition-metal complexes on the oxidation potential for $HO^-$ in MeCN illustrates some of these effects; the results are summarized in Table 8-2.[4]

In organic solvents the $HO\cdot$ radical that is produced by an SET step also may be captured by the solvent or dimerize to form hydrogen peroxide, HOOH. Bona fide examples of the latter reaction are rare because the competing reaction with solvent is faster, except perhaps in MeCN.[12-14] Also, hydrogen peroxide is highly reactive in basic solutions of dipolar aprotic solvents, including MeCN.[15-17] Thus, hydrogen peroxide may not be found, even if it is produced at some stage of the reaction. For systems in which hydrogen peroxide is suspected, the formation of dioxygen or oxygenated products may be used as indirect evidence of its intermediacy.

## Reaction classifications (single-electron shift mechanism)

The reaction continuum for $HO^-$ can be subdivided into three discrete categories that are outlined in Scheme 8-1: (1) Displacement reactions in which the leaving group departs with an electron supplied by $HO^-$ (polar-group transfer), (2) addition reactions in which a covalent bond is formed (polar-group addition), and (3) simple electron-transfer reactions in which $HO^-$ acts as an electron donor

Table 8-1  Redox Potentials for the Single-Electron (a) Oxidation of HO⁻ and Other Oxy Anion Bases and (b) Reduction of Electrophilic Substrates in Water and in Acetonitrile

a. Oxidation of Oxy-Anion Bases

| Base (:$B^-$) | $(pK_{HB})H_2O^a$ | $(E^o{}_B)H_2O^b$ (V versus NHE) | $(pK_{HB})MeCN^{a,c}$ | $(E^o{}_B)MeCN^d$ (V versus NHE) |
|---|---|---|---|---|
| $Cl^- \rightarrow Cl\cdot + e^-$ | | +2.41 | | +2.24 |
| $HO^- \rightarrow HO\cdot + e^-$ | 15.7 | +1.89 | 30.4 | +0.92 |
| $PhO^- \rightarrow PhO\cdot + e^-$ | 9.2 | ~+0.7 | 16.0 | +0.30 |
| $O_2{}^{\cdot-} \rightarrow \cdot O_2 + e^-$ | 4.9 | $-0.16^e$ | ~13 | $-0.63^e$ |
| $HOO^- \rightarrow HOO\cdot + e^-$ | 11.8 | +0.20 | ~22 | $-0.34$ |
| $H^- \rightarrow H\cdot$ | | $-2.20$ | | $-3.2$ |
| $[e^-(H_2O) \equiv H\cdot + HO^-] \rightarrow H_2O + e^-$ | | $-2.93$ | | $-3.90$ |

## b. Reduction of Electrophiles

| Electrophile (YX) | $(-\Delta G)_{Y\text{-OH}}$ (kcal)[f] | $(E^{o\prime}{}_E)_{H_2O}$ (V versus NHE) | $(E^{o\prime}{}_E)$MeCN (V vs NHE) |
|---|---|---|---|
| $Au^+ + e^- \rightarrow Au\cdot$ | 31 | +1.7 | +1.58 |
| $Ag^+ + e^- \rightarrow Ag\cdot$ | 24 | +0.80 | +0.54 |
| $(TPP)Fe^{III}(py)_2^+ + e^- \rightarrow (TPP)Fe^{II}(py)_2$ | 31 | | +0.38 |
| $MV^{2+} + e^- \rightarrow MV^+\cdot$ | 72 | -0.45 | -0.18 |
| $AQ + e^- \rightarrow AQ^-\cdot$ | 72 | | -0.58 |
| $CCl_4 + e^- \rightarrow Cl_3C\cdot + Cl^-$ | 80 | | -0.91 |
| $C_6Cl_6 + e^- \rightarrow \cdot C_6Cl_5 + Cl^-$ | 84 | | -1.26 |
| $BuCl + e^- \rightarrow Bu\cdot + Cl^-$ | 83 | | -2.5 |
| $H_3O^+ + e^- \rightarrow H\cdot + H_2O$ | 111 | -2.10 | -1.58 |

[a] $pK_a$ of the conjugate acid, Ref. 5.
[b] Ref. 2.
[c] Ref. 10.
[d] Chapter 2 and Ref. 4.
[e] Standard state for $O_2$ is 1 M.
[f] Chapter 3 and Ref. 3.

Table 8-2  Oxidation Potentials for $HO^-$ in $H_2O$ and MeCN, and in the Presence of Metal Complexes

| | $E^{o'}$ (V versus NHE) | |
|---|---|---|
| a.  Free base | $H_2O$ | MeCN |
| $HO^- \rightarrow HO\cdot + e^-$ | +1.89 | +0.92 |
| $2\ HO^- \rightarrow O^-\cdot + H_2O + e^-$ | +1.77 | +0.59 |
| $2\ HO^- \rightarrow O(g) + H_2O + 2\ e^-$ | +1.60 | +0.63 |
| $3\ HO^- \rightarrow HOO^- + H_2O + 2\ e^-$ | +0.87 | -0.11 |
| | | |
| b.  Metal–porphyrin complexes | | |
| $(TPP)Zn^{II}(^-OH)^- + HO^- \rightarrow (TPP)Zn^{II}(O\cdot\cdot)^- + H_2O + e^-$ | | +0.73 |
| $(TPP)Co^{II}(^-OH)^- \rightarrow (TPP)Co^{III}\text{-}OH + e^-$ | | +0.02 |
| $(TPP)Fe^{II}(^-OH)^- \rightarrow (TPP)Fe^{III}\text{-}OH + e^-$ | | -0.48 |
| $(TPP)Mn^{II}(^-OH)^- \rightarrow (TPP)Mn^{III}\text{-}OH + e^-$ | | -0.35 |
| | | |
| c.  Metal–$(Ph_3PO)$ complexes | | |
| $(Ph_3PO)_4Zn^{II}(OH)_2 + HO^- \rightarrow (Ph_3PO)_4Zn^{II}(OH)(O\cdot) + H_2O + e^-$ | | +0.67 |
| $(Ph_3PO)_4Ni^{II}(OH)_2 + HO^- \rightarrow (Ph_3PO)_4Ni^{III}(O)(OH) + H_2O + e^-$ | | -0.01 |
| $(Ph_3PO)_4Co^{II}(OH)_2 + HO^- \rightarrow (Ph_3PO)_4Co^{III}(O)(OH) + H_2O + e^-$ | | -0.05 |
| $(Ph_3PO)_4Fe^{II}(OH)_2 + HO^- \rightarrow (Ph_3PO)_4Fe^{III}(O)(OH) + H_2O + e^-$ | | +0.12 |
| $(Ph_3PO)_4Mn^{II}(OH)_2 + HO^- \rightarrow (Ph_3PO)_4Mn^{III}(O)(OH) + H_2O + e^-$ | | +0.31 |

Scheme 8-1  Single-Electron Shift (Equivalent to the Transfer of an Electron from O to X)

$$O \rightarrow X$$
single-electron shift

a.  Polar-group transfer (e.g., a Brønsted proton-transfer reaction or a nucleophilic-displacement reaction).

$$HO\colon^- + Y..X \rightarrow HO\cdot\cdot Y + \colon X^-$$

b.  Polar-group coupling (e.g., a nucleophilic-addition reaction to a double bond; $X=C$, $Y=O$).

$$HO\colon^- + Y\text{-}X \rightarrow HO\cdot\cdot Y\text{-}\ddot{X}$$

c.  Single-electron transfer (SET)

$$HO\colon^- + Y\text{-}X \rightarrow HO\cdot + (Y\text{-}X)^-\cdot$$

(single-electron transfer). This view of the chemistry of $HO^-$ also applies to the reactions of superoxide ion ($O_2^{-\cdot}$) and other nucleophilic oxy anions (Table 8-1a).

The polar pathways are formally equivalent to a discrete electron-transfer step, that is, a pure SET step that is followed by a chemical step. If a hypothetical SET step is followed by coupling of a free-radical pair that is produced in the SET step, the overall reaction is the equivalent of a polar-group-coupling reaction [Reaction (b) of Scheme 8-1]. If the coupling is accompanied by the elimination of a leaving group, a polar-group-transfer reaction results [Reaction (a) of Scheme 8-1].

The "single-electron shift" mechanism appears to be general and applicable for electron-, proton-, atom-, and group-transfer reactions. The assumptions for this proposition include: (1) polar and SET pathways share a common feature, a single-electron shift from an electron donor to an electron acceptor; (2) the barrier heights for the exchange of an electron, proton, atom, or group of atoms can be described by the same general equation[18]; (3) for an unsymmetrical polar-group-transfer reaction the net energy change, $-\Delta G$ (the thermodynamic component of the energy barrier), is proportional to the sum of (a) the difference of the redox potentials for the electron acceptor $(Y-X)$ and the electron donor $(HO:^-)$ and (b) the bond energy (BE) for the group transfer product $(HO-Y;$ see Scheme 8-2).

*Scheme 8-2 Redox Energetics for Single-Electron-Shift Reactions*

Energy change(J)
_____

*Electron donor:* $\qquad HO:^- \rightarrow HO\cdot + e^-$ $\qquad\qquad -(E°'_{HO\cdot/HO^-})nF$

*Electron acceptor:* $\quad Y-X + e^- \rightarrow Y-X^{-\cdot} \rightarrow Y\cdot + X:^-$ $\quad (E°'_{XY/XY^-})nF$

*Bond formation:* $\qquad HO\cdot + Y\cdot \rightarrow HO-Y$ $\qquad\qquad (-\Delta G_{BF})_{HO-Y}nF/23.1$

*Polar-group transfer:*

$\quad HO:^- + Y-X \rightarrow HO-Y + X:^-$ $\qquad -\Delta G_{reac} = nF[(E°'_{YX/YX^-} - E°'_{HO\cdot/HO^-})$
$\qquad\qquad\qquad\qquad\qquad\qquad\qquad\qquad\qquad\qquad + (-\Delta G_{BF})_{HO-X}/23.1]$
$\qquad\qquad\qquad\qquad\qquad\qquad\qquad\qquad\qquad = nF[\Delta E°'_{reac} + (-\Delta G_{BF})_{HO-X}/23.1]$

*Polar-group Coupling:*

$\quad HO:^- + Y-X \rightarrow HO:Y:X^-$ $\qquad -\Delta G_{reac} = nF[\Delta E°'_{reac} + (-\Delta G_{BF})_{HO-YX^-}/23.1]$

*Single-electron transfer (SET):*

$\quad HO:^- + Y-X \rightarrow HO\cdot + Y-X^{-\cdot}$ $\qquad -\Delta G_{reac} = nF\ \Delta E°'_{reac}$

Table 8-3 lists a number of reactions of hydroxide ion and, for comparison, superoxide ion with electron donors. These reactions are classified conventionally and according to the categories shown in Scheme 8-1. When water is replaced by a dipolar aprotic solvent [e.g., acetonitrile (MeCN) or dimethyl sulfoxide (Me$_2$SO)], many of these electron donor-electron acceptor reactions proceed at dramatically faster rates and produce much larger yields of free-radical products. [Dimethylformamide (DMF) is generally avoided because of the possible hydrolysis of the amide bond by hydroxide ion.][19]

**Table 8-3  Reactions of Hydroxide Ion and Superoxide Ion with Electrophiles**

1. Polar-group-transfer reactions
   a. Deprotonation by hydroxide ion
      $HO^- + H_3O^+ \rightarrow 2\ H_2O$
      $HO^- + CH_3OH \rightarrow H_2O + CH_3O^-$
      $HO^- + HOC(O)R \rightarrow H_2O + {}^-OC(O)R$
      $O_2^{-\cdot} + NH_4^+ \rightarrow HOO\cdot + NH_3$
      $O_2^{-\cdot} + HOCH_3 \rightarrow HOO\cdot + CH_3O^-$

   b. Nucleophilic-substitution reactions
      $HO^- + RCl \rightarrow R\text{-}OH + Cl^-$
      $O_2^{-\cdot} + RCl + ROO\cdot + Cl^-$
      $HO^- + RC(O)OR' \rightarrow RC(O)OH + R'O^- \rightarrow RC(O)O^- + R'OH$
      $O_2^{-\cdot} + RC(O)OR' \rightarrow RC(O)OO\cdot + RO^-$
      $4\ HO^- + CCl_4 \rightarrow 3\ Cl^- + ClC(O)O^- + 2\ H_2O$
      $5\ O_2^{-\cdot} + CCl_4 + H_2O \rightarrow 4\ Cl^- + HOC(O)O^- + \frac{7}{2}O_2 + \frac{1}{2}HOOH$
      $HO^- + C_6Cl_6 \rightarrow C_6Cl_5OH + Cl^-$

2. Polar-group-coupling reactions.
   a. Nucleophilic-addition reactions
      $HO^- + CO_2 \rightarrow HOC(O)O^-$
      $O_2^{-\cdot} + CO_2 \rightarrow \cdot OOC(O)O^-$

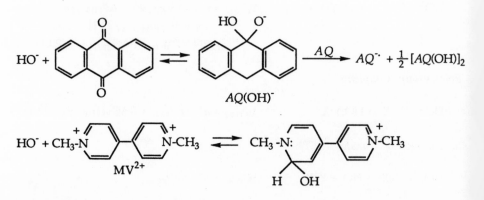

Table 8-3 (cont.)

b. Reductive-addition reactions
   $HO^- + Au^+ \rightarrow Au\text{-}OH$
   $HO^- + Fe^{III}(TPP)^+ \rightarrow (TPP)Fe^{III}\text{-}OH$
   $HO^- + Fe^{III}(TPP)(py)_2^+ \rightarrow Fe^{II}(TPP) + (1/n)\,[py(\cdot OH)]_n$
   Other $M(III)/^-OH$ reactions

3. Single-electron-transfer reactions

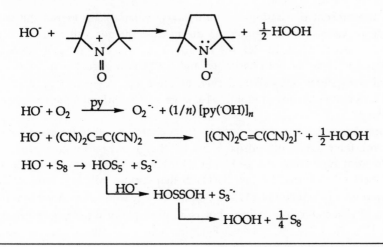

$HO^- + O_2 \xrightarrow{\;py\;} O_2^{-\cdot} + (1/n)\,[py(^\cdot OH)]_n$

$HO^- + (CN)_2C{=}C(CN)_2 \longrightarrow [(CN)_2C{=}C(CN)_2]^{-\cdot} + \tfrac{1}{2}HOOH$

$HO^- + S_8 \rightarrow HOS_5^\cdot + S_3^{-\cdot}$

$\qquad\quad \underline{\big| HO^-} \rightarrow HOSSOH + S_3^{-\cdot}$

$\qquad\qquad\quad \big\lrcorner \longrightarrow HOOH + \tfrac{1}{4}S_8$

---

As the prototype reactions in Scheme 8-1 imply, a reaction that involves a single-electron shift may not produce observable free-radical products. Conversely, the failure to find free-radical products does not prove the absence of a single-electron-shift mechanism. Other arguments are necessary to establish the nature of polar-group-transfer and polar-coupling reactions.

*Polar-group-transfer reactions*

*Deprotonation by hydroxide ion.* Although Brønsted proton-transfer reactions appear to belong to a unique category not described by Scheme 8-1, they are examples of polar-group-transfer reactions and are not different in principle from nucleophilic-displacement reactions. Deprotonation by hydroxide ion can be regarded as the shift of an electron from $HO^-$ to the Brønsted acid synchronously with the transfer of a hydrogen atom from the Brønsted acid to the incipient $HO\cdot$ radical, with the reaction driven by covalent-bond formation between the $HO\cdot$ radical and the $H\cdot$ atom to form water.

$$HO\!:^- + H\!\cdot\!B \rightleftharpoons [HO\cdot\!\cdot H\!:\!B^-] \rightleftharpoons HO\!:\!H + B\!:^-, \quad (-\Delta G_{BF})_{HO-H},\ 111\ kcal \quad (8\text{-}3)$$

Consistent with the idea that the single-electron shift is a fundamental process is the notion that the electron distribution (or partial charge) on atoms near the reaction site changes during the course of the reaction. This is so obvious as to seem trivial, but bears repeating because our thinking often is misdirected by the assignment of oxidation numbers (or oxidation states) to atoms via arbitrary rules. For example, by convention the oxidation state +1 assigned to the hydrogen and -2 is assigned to the oxygen in hydronium ion, water, and hydroxide ion. However, the partial charges on the hydrogen and oxygen atoms change substantially in this series: $H_3O^+$(+0.35, -0.05); $H_2O$(+0.12, -0.25); and $HO^-$ (-0.35, -0.65).[20]

In the present discussion the term *charge number* represents the partial charge on an atom, and is defined as the nearest integer value of the partial charge on an atom. Thus, in $HO^-$ the hydrogen is "zero charge" and the oxygen "-1 charge," and in water both hydrogen and oxygen are zero charge. For $H_3O^+$ the partial charge that is distributed over the three equivalent hydrogen atoms is assigned to a single hydrogen atom, (+1-charge hydrogen), and the oxygen atom is zero charge.

From Scheme 8-1 and the notion of the charge number (or partial charge on an atom), the reaction of hydroxide ion (-1-charge oxygen) with a proton (+1-charge) to form $H_2O$ (zero-charge H and O) involves a single-electron shift from $HO^-$ to $H^+(H_2O)$ to form $HO\cdot$ and $\cdot H(H_2O)$ that form a covalent bond (HO–H) with a bond energy ($-\Delta G_{BF}$) of 111 kcal/mol.[3] This reaction does not result in the formation of detectable free radical, but is an example of a polar-group-transfer reaction [Reaction (b) of Scheme 8-1] in which an atom transfer occurs synchronously with the electron shift.

Thus, the reaction of $H^+$ with $HO^-$ is a prototype example of a charge-transfer or redox reaction that is also a polar-group-transfer reaction. It can be resolved into three component reactions (see Scheme 8-2):

$$H^+(aq) + e^- \rightarrow H\cdot(aq), \quad E^{o'}_{8\text{-}4}, \text{-2.10 V versus NHE} \qquad (8\text{-}4)$$

$$HO\cdot(aq) + e^- \rightarrow HO^-(aq), \quad E^{o'}_{8\text{-}5}, +1.89 \text{ V} \qquad (8\text{-}5)$$

$$H\cdot(g) + HO\cdot(g) \rightarrow HOH(g), \quad -\Delta G^{o'}_{8\text{-}6}, 111.4 \text{ kcal } (+4.83 \text{ V}) \qquad (8\text{-}6)$$

These can be combined [Eq. (8-4) - Eq. (8-5) + Eq. (8-6)] to give

$$H^+(aq) + HO^-(aq) \xrightarrow{K_f} HOH(aq), \quad -\Delta G^{o'}_{8\text{-}7} = nE^{o'}_{8\text{-}7}\, F \qquad (8\text{-}7)$$

$$-\Delta G^{o'}_{8\text{-}7} = -(\Delta G^{o'}_{8\text{-}4} - \Delta G^{o'}_{8\text{-}5} + \Delta G^{o'}_{8\text{-}6}) = 19.2 \text{ kcal } (+0.83 \text{ V}) \qquad (8\text{-}8)$$

The evaluation of $-\Delta G_{8-7}^{o\prime}$ neglects the small differences in the hydration energies of H·, HO·, and $H_2O$, but provides a reasonable measure of the formation constant ($K_f$)

$$\log K_f = \frac{+0.83}{0.059} = 14 \tag{8-9}$$

In the net reaction [Eq. (8-7)] the energy from bond formation [Eq. (8-6)] provides a driving force of nearly 5 V, which is more than sufficient to overcome the unfavorable electron-transfer energy (-3.89 V).

The reaction of superoxide ion ($O_2^{-}\cdot$), a radical anion, with water also can be viewed as a polar-group-transfer reaction.

$$\cdot OO\overset{\frown}{:} + H\colon\!OH \rightleftharpoons [\cdot OO\colon\!H\colon^-OH] \rightleftharpoons \cdot OO\colon\!H + {}^-\colon\!OH \tag{8-10}$$

Here, the product HOO· is a free radical that reacts bimolecularly to form hydrogen peroxide and dioxygen.

$$2\,HOO\cdot \rightarrow HOOH + O_2 \tag{8-11}$$

The formation of the stable covalent bonds in the product molecules provides the driving force that allows superoxide ion to deprotonate Brønsted acids that are much weaker acids than HOO·.

Because every chemical reaction involves charge transfer (or at least partial electron shifts), the distinction between an acid–base reaction and an oxidation–reduction reaction becomes meaningless unless defined in terms of changes in conventionally assigned oxidation number.[21] This point of view also has been expressed before, but still is not discussed in contemporary textbooks of general, organic, and inorganic chemistry.

*Nucleophilic substitution reactions.* The view that substitution or displacement reactions that involve hydroxide ion are examples of polar-group-transfer reactions (with a single-electron shift) is probably the least iconoclastic proposal. Most accept the view that many nucleophilic displacement reactions occur by a SET mechanism.[22] In a number of cases free-radical intermediates have been identified, which is consistent with a discrete SET step. Only a slight extension of this concept is required to encompass all nucleophilic reactions within the categories described in Scheme 8-1.

The reactions of HO⁻ and $O_2^{-}\cdot$ with alkyl halides exhibit the same general pattern (Scheme 8-3), with second-order kinetics and inversion of configuration.[23,34] Free radicals are not detected in the reactions with hydroxide ion, which indicates that there probably is not a discrete SET step, but rather that the transfer of the entering and leaving groups is synchronous with a single-electron shift.

*Scheme 8-3   Nucleophilic substitution*

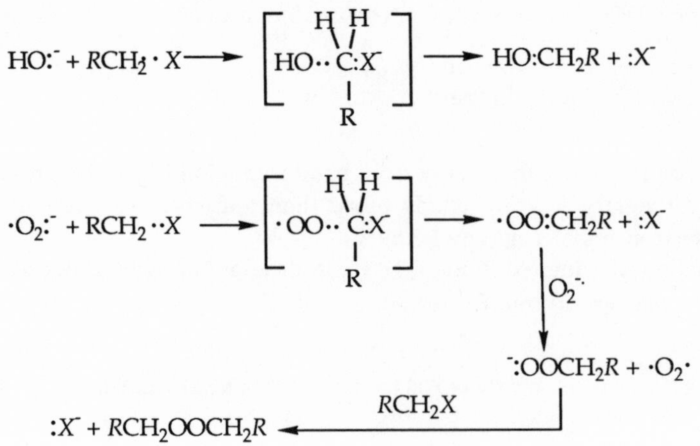

$$HO:^- + RCH_2 \cdot X \longrightarrow \left[ \begin{array}{c} H \; H \\ \backslash \; / \\ HO \cdots C:X \\ | \\ R \end{array} \right] \longrightarrow HO:CH_2R + :X^-$$

$$\cdot O_2:^- + RCH_2 \cdots X \longrightarrow \left[ \begin{array}{c} H \; H \\ \backslash \; / \\ \cdot OO \cdots C:X \\ | \\ R \end{array} \right] \longrightarrow \cdot OO:CH_2R + :X^-$$

$$:X^- + RCH_2OOCH_2R \xleftarrow{\quad RCH_2X \quad} \quad ^-:OOCH_2R + \cdot O_2\cdot$$

(with $\Big| O_2^{-\cdot}$ step leading to $^-:OOCH_2R + \cdot O_2\cdot$)

The reaction of superoxide ion with alkyl halides produces a free radical in the primary step, because the spin angular momentum of the unpaired electron of the superoxide anion radical is conserved. Again, the transfer of the entering and leaving group is presumed to be synchronous with the single-electron shift. The alternative mechanism, a discrete SET step followed by expulsion of the leaving group and coupling of the radical with dioxygen, is implausible because the reduction potentials of most alkyl halides are at least 0.8 V more negative than the $O_2/O_2^{-\cdot}$ redox potential and the formation of alkyl radicals has not been observed.

Thus, the reaction of $HO^-$ with BuBr is another example of a redox reaction that involves a polar-group transfer. This can be resolved into three component reactions (in MeCN)

$$BuBr + e^- \rightarrow Bu\cdot + Br^-, \quad E^{\circ\prime}_{8\text{-}12}, -1.45 \text{ V versus NHE} \qquad (8\text{-}12)$$

$$HO\cdot + e^- \rightarrow HO^-, \quad E^{\circ\prime}_{8\text{-}13}, +1.89 \text{ V} \qquad (8\text{-}13)$$

$$Bu\cdot(g) + HO\cdot(g) \rightarrow BuOH(g), \quad -\Delta G^{\circ\prime}_{8\text{-}14}, 86 \text{ kcal } (+3.72 \text{ V}) \qquad (8\text{-}14)$$

These can be combined [Eq. (8-12) - Eq. (8-13) + Eq. (8-14) to give

$$BuBr + HO^- \rightarrow BuOH + Br^-, -\Delta G^{\circ\prime}_{8\text{-}15} = nE^{\circ\prime}_{8\text{-}15}F \qquad (8\text{-}15)$$

$$-\Delta G^{\circ\prime}_{8\text{-}15} = -(\Delta G^{\circ\prime}_{8\text{-}12} - \Delta G^{\circ\prime}_{8\text{-}13} + \Delta G^{\circ\prime}_{8\text{-}14}) = 8.8 \text{ kcal } (+0.38 \text{ V}) \qquad (8\text{-}16)$$

Similar analyses are possible for the initial polar-group transfer step for $CCl_4$ ($E^{\circ\prime}_{red}$, -0.91 V versus NHE), $C_6Cl_6$ ($E^{\circ\prime}_{red}$, -1.26 V), and $C_{12}Cl_{10}$ (PCB; $E^{\circ\prime}_{red}$, -1.30 V). Each of these substrates undergoes a net exergonic redox reaction with $HO^-$; the initial step is analogous to that of Eq. (8-15).

The reactions of $CCl_4$ with $HO^-$ and $O_2^{-\cdot}$ are complex multistep reactions and the nature of the primary step is not well understood. The initial reaction is followed by even faster secondary reactions that ultimately result in the almost complete oxygenation of the carbon and the release of the chlorine as chloride ion. The overall reaction for superoxide ion with $CCl_4$ (followed by dilution with water),

$$5\,O_2^{-\cdot} + CCl_4 + H_2O \;\rightarrow\; HOC(O)O^- + 4\,Cl^- + \tfrac{7}{2}O_2 + \tfrac{1}{2}HOOH \tag{8-17}$$

results in the stoichiometric formation of bicarbonate ion.[25] The reaction of hydroxide ion with $CCl_4$ in dimethyl sulfoxide (followed by dilution with water) has a net stoichiometry that is consistent with the reaction

$$4\,HO^- + CCl_4 \;\rightarrow\; ClC(O)O^- + 3\,Cl^- + 2\,H_2O \tag{8-18}$$

(without positive identification of chloroformate ion).[26]

At least three alternatives must be considered to understand the nature of the primary reaction between $HO^-$ and $CCl_4$ (see Scheme 8-4): (1) A discrete SET reaction; (2) A polar-group transfer on carbon, that is, the formation of an HO–C bond with the concerted displacement of $Cl^-$ (nucleophilic attack on carbon); (3) A polar-group transfer on chlorine, that is, the formation of an HO–Cl bond with the concerted displacement of $Cl_3C:^-$ (nuleophilic attack on chlorine).

*Scheme 8-4 Nucleophilic addition/displacement*

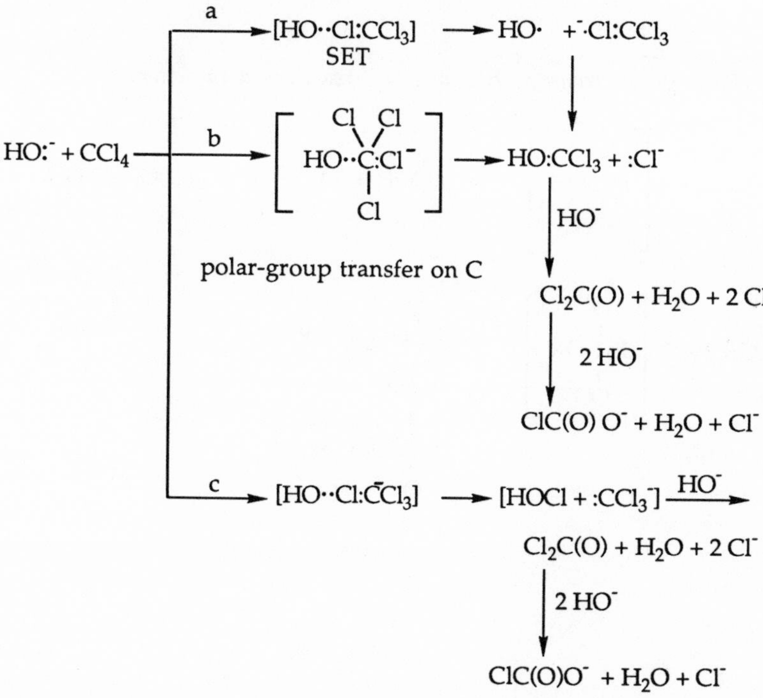

The first alternative, a single-electron transfer from $HO^-$ to $CCl_4$ is improbable because the reaction is about 1.5 V more endergonic than the analogous reaction with superoxide ion, yet proceeds at about 70% of the rate with $O_2^{-\cdot}$.[27] The rate of reaction of $O_2^{-\cdot}$ with polyhalogenated compounds $RCCl_3$ decreases about $10^5$ per volt change in $E^{o\prime}$.[25] Although Reaction (b) of Scheme 8-4 would be expected to have a large reaction barrier due to the substantial atom motion required to enable a polar-group transfer on carbon, a recent study of the reaction of $O_2^{-\cdot}$ with $CCl_4$ indicates that $Cl_3COO\cdot$ is the primary product [probably via Reaction (a) of Scheme 8-4].[28] A third possibility for the reaction of $HO^-$ may be a polar-group transfer on chlorine [Reaction (c) of Scheme 8-4], which is analogous to a mechanism proposed for the reaction of $CCl_4$ with potassium *t*-butoxide.[29] For the latter, *t*-BuOCl is proposed as an intermediate to explain the isobutylene oxide product.

$$t\text{-BuOCl} + t\text{-BuO}^- \rightarrow (CH_3)_2\overset{\displaystyle}{\underset{\displaystyle O}{C\text{-}CH_2}} + Cl^- + t\text{-BuOH} \qquad (8\text{-}19)$$

Finally, there is the possibility that the primary reaction involves another species entirely, such as an anion produced by deprotonating the solvent or from the addition of $HO^-$ to a solvent molecule.

The reactions of $HO^-$ and $O_2^{-\cdot}$ with the carbonyl group of esters[30,31] and quinones[32,33] share a common feature, the addition of the nucleophile to the carbonyl carbon. When a suitable leaving group is present, there is an essentially concerted elimination to give products, as shown in Scheme 8-5. The addition of $HO^-$ to a carbonyl group in quinones, which have no leaving group, yields an adduct that is sufficiently stable to characterize.[32]

*Scheme 8-5 Polar-group transfer; $HO^-$ and $O_2^{-\cdot}$ reactions with esters*

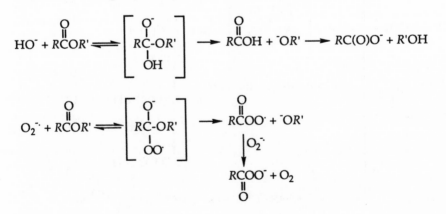

## Polar-group-coupling reactions

*Nucleophilic-addition reactions.* The most common addition reaction is to a carbonyl group without an adequate leaving group. Examples include the reaction of $HO^-$ and $O_2^{-\cdot}$ with $CO_2$ and quinones.[32] A common feature of these reactions is the formation of an adduct that is sufficiently stable to be isolated or characterized (see Scheme 8-6). The same orange-colored species results from the reaction of solid tetramethylammonium superoxide with gaseous $CO_2$ and with neat $CCl_4$,[34] and is believed to be an anion radical, $\cdot OOC(O)O^-$.[28]

*Scheme 8-6  Polar-group coupling*

a. $HO^-$ and $O_2^{-\cdot}$ reactions with $CO_2$.

$$HO^- + CO_2 \rightleftharpoons HOC(O)O^-$$

$$O_2^{-\cdot} + CO_2 \rightleftharpoons \cdot OOC(O)O^-$$

b. $HO^-$ reaction with 9,10-anthraquinone.

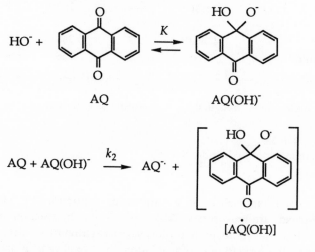

$$\tfrac{1}{2}[AQ(OH)]_2 \longrightarrow AQ + \tfrac{1}{2}HOOH$$

c. $HO^-$ reaction with benzylviologen ($BV^{2+}$).

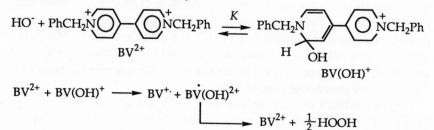

$$BV^{2+} + BV(OH)^+ \longrightarrow BV^{+\cdot} + BV(OH)^{2+}$$

$$\longrightarrow BV^{2+} + \tfrac{1}{2}HOOH$$

The reaction of HO⁻ with 9,10-anthraquinone in MeCN produces an adduct (stable at -20°C),[32] which reacts further at room temperature to yield the semiquinone anion radical (AQ⁻·). The equilibrium constants for the formation of the adducts and the rate constants for the reaction of the adduct with a second quinone molecule are given in Table 8-4 (see Scheme 8-6).

Table 8-4  Equilibrium and Rate Constants for the Reactions of Quinones with Hydroxide Ion (see Scheme 8-6)[a]

| Quinone | $K_1$ ($M^{-1}$) | $k_2$ ($M^{-1}s^{-1}$) | Q(OH)⁻ adduct $\lambda_{max}$ (nm) | log ε |
|---|---|---|---|---|
| *p*-benzoquinone monosulfonate | 125 | — | 253 | 3.86 |
| Chloro-*p*-benzoquinone | 364 | — | 267 | 4.00 |
| 2,5-dichloro-*p*-benzoquinone | $1.7 \times 10^3$ | — | 250 | 4.00 |
| Trichlorohydroxy-*p*-benzoquinone | 210 | — | 250 | 3.75 |
|  |  |  | 365 | 3.70 |
| Tetrachloro-p-benzoquinone[b] (Chloranil) | $1 \times 10^5$ | — | 285 | 3.83 |
| 9,10-anthraquinone[c] | $4.3 \times 10^4$ | 1.2 | 268 | 4.3 |
| 2-ethyl-9,10-anthraquinone[c] | $4 \times 10^4$ | 4.2 | 268 | — |

[a] Aqueous phosphate buffer, ionic strength 0.375, Ref. 33 (unless otherwise indicated).

[b] 50% aqueous ethanol, Ref. 33.

[c] MeCN/tetra-*n*-butylammonium hydroxide, Ref. 32.

*Reductive-addition reactions.* Several examples of reduction by HO⁻ of transition-metal complexes are known (see Table 8-3).[4,35–39] The reaction of Au⁺ with HO⁻ in MeCN is believed to be a prototype of reactions that involve a single-electron shift and the formation of a metal atom/hydroxyl radical bond.

$$\text{HO:}^- + \text{Au}^+ \rightarrow [\text{HO} \cdot \cdot \text{Au}] \rightarrow \text{HO:Au}^I(s) \qquad (8\text{-}20)$$

The $E^{\circ\prime}$ for the Au⁺/Au redox couple in MeCN is +1.58 V versus NHE compared to +0.9 V for the HO·/HO⁻ couple;[4,35] hence electron transfer is an exergonic process. Electrochemical oxidation of HO⁻ at a gold electrode in MeCN occurs at -0.19 V versus NHE, which indicates an Au–OH bond energy of 26 kcal (-$\Delta G_{BF}$). [However, a Au–Au bond must be broken (54 kcal); this gives a value of 53 kcal (26 + 54/2) for the Au-OH bond].[3]

The addition of HO⁻ to manganese(III) complexes [$Mn^{III}(O_2\text{-bpy})_3{}^{3+}$, $Mn^{III}(TPP)^+$ (TPP = tetraphenylporphyrin), $Mn^{III}(PA)_3$ (PA=picolinate), and

$Mn^{III}(OAc)_3$ in aprotic media results in the rapid precipitation of the same inorganic manganese oxide.[4]

$$Mn^{III}L_3 + 3\,HO^- \rightarrow 3\,L^- + Mn^{III}(OH)_3(s) + H_2O \tag{8-21}$$

The ligands of $Mn^{III}L_3$ are reduced by electron transfer from three $HO^-$ ions. The resulting $HO\cdot$ radicals are stabilized via three $d^5sp\text{-}p$ covalent metal–oxygen bonds. Similar electron-transfer reductions by $HO^-$ have been reported for Ru(III) complexes[36] and for $(py)_2Fe^{III}TPP(ClO_4)$.[37] A mechanistic pathway for the latter process is outlined in Scheme 8-7, as is the reduction of $(TPP)Mn^{III}(ClO_4)$ by $HO^-$.[4] The direct reduction of $X$ within $[(TPP)Fe^{III}X]$ via addition of $HO^-$ has been demonstrated by electrochemical measurements[38] and an NMR titration (Scheme 8-7).[39]

*Scheme 8-7 Reductive addition*

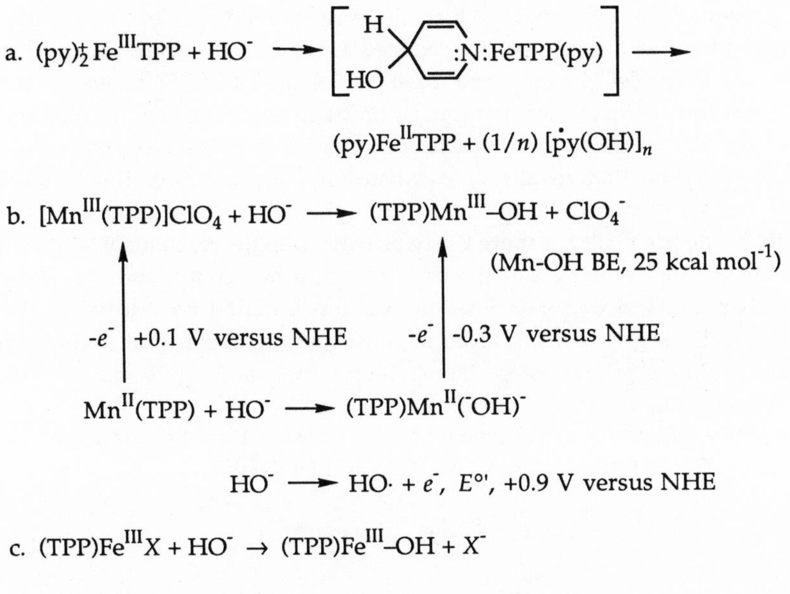

a. $(py)_2^+ Fe^{III}TPP + HO^- \longrightarrow \left[ \begin{array}{c} H \\ \diagdown \\ HO \end{array} \diagup\!\!\!\diagup :N{:}FeTPP(py) \right] \longrightarrow$

$$(py)Fe^{II}TPP + (1/n)\,[\dot{p}y(OH)]_n$$

b. $[Mn^{III}(TPP)]ClO_4 + HO^- \longrightarrow (TPP)Mn^{III}\text{–}OH + ClO_4^-$

(Mn-OH BE, 25 kcal mol$^{-1}$)

$-e^- \mid +0.1$ V versus NHE     $-e^- \mid -0.3$ V versus NHE

$Mn^{II}(TPP) + HO^- \longrightarrow (TPP)Mn^{II}(\bar{\,}OH)^-$

$HO^- \longrightarrow HO\cdot + e^-,\ E^{o\prime}, +0.9$ V versus NHE

c. $(TPP)Fe^{III}X + HO^- \rightarrow (TPP)Fe^{III}\text{–}OH + X^-$

Another example of reductive addition by $HO^-$ is its termolecular reaction with reduced $Fe^{II}(TPP)$ in the presence of dioxygen[31]

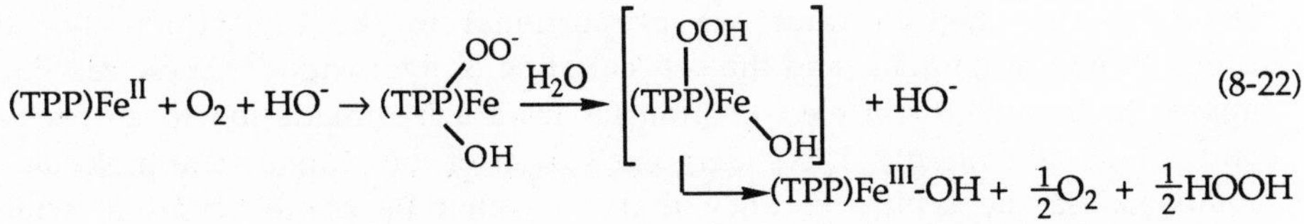

$$(TPP)Fe^{II} + O_2 + HO^- \rightarrow (TPP)Fe \overset{OO^-}{\underset{OH}{\diagup}} \xrightarrow{H_2O} \left[ (TPP)Fe \overset{OOH}{\underset{OH}{\diagup}} \right] + HO^- \tag{8-22}$$

$$\longrightarrow (TPP)Fe^{III}\text{-}OH + \tfrac{1}{2}O_2 + \tfrac{1}{2}HOOH$$

The net effect is the reduction of $O_2$ to bound superoxide ion, which hydrolyzes to bound $HOO\cdot$; the latter dissociates from the iron center and disproportionates to $HOOH$ and $O_2$.

## Single-electron-transfer reactions

The most striking and unexpected reactions of $HO^-$ are those that produce anion radicals when hydroxide is added to solutions of aromatic ketones, quinones, paraquats, and strong electron acceptors such as tetracyanoethene.[22] If the primary reaction is a SET reaction, a free-radical pair will be produced

$$HO:^- + Y:X \rightarrow HO\cdot + Y:X^{-\cdot} \tag{8-23}$$

Because $HO\cdot$ is a strong oxidant (although about 1 V weaker in MeCN than in water), a single-electron transfer would be greater than 1 V endergonic for quinones, paraquats, and ketones. All of these compounds, unless substituted with electron-withdrawing groups, are reduced at potentials less negative than -0.3 V versus NHE in MeCN, compared to +0.9 V for the $HO\cdot/HO^-$ redox couple. Hence, the reactions of quinones, paraquats, and ketones with $HO^-$ are unlikely examples of the SET process. Only synchronous coupling of the electron transfer to a chemical reaction that results in covalent-bond formation of the $HO\cdot$ can account for the spontaneous reactivity. Although the coupling of two $HO\cdot$ to form $HOOH$ is one possibility, a more likely primary step for reactants is attack by $HO^-$ of an unsaturated center (aromatic or carbonyl carbon compound), or, in the case of the paraquats, a dequaternization reaction initiated by addition. The observed free radicals are then produced in subsequent reactions that involve the intermediates from the primary step. Thus, the production of a free radical is not adequate evidence that the primary step is a SET reaction.

In aprotic solvents $HO^-$ reacts with elemental sulfur ($S_8$) via a net SET process to give the trisulfide anion radical ($S_3^{-\cdot}$) and $HOOH$.[40]

$$3\,S_8 + 8\,HO^- \rightarrow 8\,S_3^{-\cdot} + 4\,HOOH \tag{8-24}$$

Perhaps this is the most compelling example that $HO^-$ represents a stabilized electron and can affect the electron-transfer reduction of a nonmetal to give an anion radical.

Some of the reactions of $HO^-$ with quinones are accelerated by light ($\lambda <$ 500 nm); the reaction rates are proportional to the light intensity, the concentration of quinone, and the concentration of hydroxide.[22] These reactions appear to involve direct electron transfer from a hydroxide ion to an excited anthraquinone molecule. The endergonic barrier of the ground-state molecule is removed via the orbital vacancy that is created by photoexcitation, which corresponds to an energy state of greater electron affinity. The effect of light in the reactions of $HO^-$ with electron acceptor molecules that absorb in the uv-visible may be more general than heretofore realized.

Any bona fide example of a primary SET step must involve an electron acceptor with a positive redox potential so that the electron transfer is not strongly endergonic. Two molecules that react with HO⁻ fulfill this condition; tetracyano-ethene and 2,2,6,6-tetramethyl-piperidine-1-oxonium (TEMPO⁺, Table 8-3). What is the fate of an HO· radical that is produced in a SET reaction? Pulse radiolysis studies in aqueous solution confirm that it is highly reactive[12] and that it rapidly disappears via the paths outlined in Scheme 8-8 to give various products (these provide circumstantial evidence for its production).

*Scheme 8-8   Hydroxyl-radical reaction patterns*

*a.   H-atom abstraction from aliphatic compounds.*

$$HO· + RH \longrightarrow R· + HOH$$

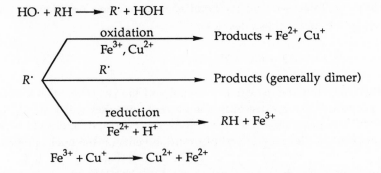

$$Fe^{3+} + Cu^+ \longrightarrow Cu^{2+} + Fe^{2+}$$

*b.   Hydroxyl-radical addition to aromatic compounds.*

$$HO· + ArH \longrightarrow ArH(OH)$$

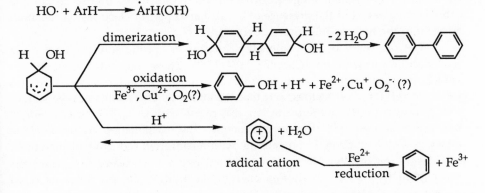

The production of a stoichiometric quantity of hydrogen peroxide (in an $O_2$-free system) provides strong evidence for the formation of HO· radical. Yields of 90% are observed from the combination of HOO⁻ and TEMPO⁺ via an iodide titration of the carbonate-buffered aqueous solution, with quantitative recovery of the TEMPO-free-radical product by ether extraction of the aqueous layer.[41] The mass spectrum of the reaction products includes a peak at mass 34.

Although this reaction appears to be a clear-cut example of a SET reaction, several presumptions are necessary. First the bimolecular reaction to form hydrogen peroxide is assumed to be faster than reaction of HO· with either $CO_3^{2-}$ or with TEMPO radical (the latter is produced with HO· in the solvent cage). Second, formation of HOOH is assumed to be solely via HO· coupling (it also can be formed from secondary reactions; e.g., from peroxy dicarbonates).[34]

Because iodometry is a nonselective method that measures other peroxides as well as hydrogen peroxide, the mass-34 peak is the only definitive evidence for HOOH.

### Relative reactivity of $HOO^-/O_2^{-\cdot}/HO^-$ with electrophilic substrates

The nucleophilicity of oxy anions $(YO^-)$ is directly related to their oxidation potentials $(E^{\circ\prime}{}_{B^-/B}$, Table 8-1) and the bond energies of their products $(YO\text{-}R$, Chapter 3) with electrophilic substrates $(RX)$

$$YO^- + RX \rightarrow YO\text{--}R + X^- \qquad (8\text{-}25)$$

Hence, the more negative the oxidation potential and the larger the $YO\text{--}R$ bond energy, the proportionally greater nucleophilic reactivity that will result. The shift in the oxidation potential of $HO^-$ from +1.89 V versus NHE in $H_2O$ to +0.92 V in MeCN reflects the "leveling effect" of protic solvents on the nucleophilicity of oxy anions. Likewise, the shift in potential for $HOO^-$ [+0.20 V ($H_2O$) to -0.34 V (MeCN)] is in accord with the exceptional reactivity of $HOO^-$ in aprotic solvents. In aqueous media the reactivity of $HOO^-$ is leveled by extensive anionic solvation,[42] but remains significant with many substrates due to its unique orbital energies[43] and the presence of an unshared pair of electrons on the atom adjacent to the nucleophilic center ($\alpha$ effect).[44]

The reactivity of $O_2^{-\cdot}$ with alkyl halides in aprotic solvents occurs via nucleophilic substitution (Chapter 7).[23-25,45] These and subsequent kinetic studies confirm that the reaction order is primary>secondary>>>tertiary and I>Br>Cl>>>F for alkyl halides, and that the attack by $O_2^{-\cdot}$ results in inversion of configuration ($S_N2$). Superoxide ion also reacts with $CCl_4$,[25,26] $Br(CH_2)_2Br$,[46] $C_6Cl_6$,[47,48] and esters[49-51] in aprotic media. The reactions are via nucleophilic attack by $O_2^{-\cdot}$ on carbon, or on chlorine with a concerted reductive displacement of chloride ion or alkoxide ion. As with all oxy anions, water suppresses the nucleophilicity of $O_2^{-\cdot}$ (hydration energy, 100 kcal)[52] and promotes its rapid hydrolysis and disproportionation. The reaction pathways for these compounds produce peroxy radical and peroxide ion intermediates ($ROO\cdot$ and $ROO^-$).

Hydroperoxide ion ($HOO^-$) is unstable in most aprotic solvents, but persists for several minutes in pyridine ($k_{decomp}$, 4.6 x $10^{-3}$ s$^{-1}$), which allows studies of its nucleophilic reactivity. In pyridine $HOO^-$ is oxidized in an one-electron transfer to give $HOO\cdot$, which is in accord with previous studies in MeCN.[53]

$$HOO^- \rightarrow HOO\cdot + e^-, \quad E_{p,a}, -0.34 \text{ V versus NHE (MeCN)} \qquad (8\text{-}26)$$

In the case of HOO⁻ (and $t$-BuOO⁻), the general "leveling" of nucleophilic reactivity by protic solvents (water and alcohols) enhances its lifetime such that the net reactions for HOO⁻ with electrophilic substrates usually are most efficient and complete in $H_2O$ or $t$-BuOOH. Almost all other solvents react with HOO⁻ or facilitate its decomposition. The relative lifetime of HOO⁻ in various solvents is in the order: $H_2O$>>MeOH>EtOH>>diglyme>pyridine~PEGM 350>>MeCN> DMF>>Me$_2$SO.[54]

Table 8-5 summarizes the relative reactivities of HO⁻($H_2O$)/HO⁻ (MeOH)/HOO⁻/O$_2$⁻· with halocarbons and esters in pyridine and the kinetics for the reaction of O$_2$⁻· with the substrates in aprotic solvents.[26,54] The relative reactivity with primary halides ($n$-BuBr and PhCH$_2$Br) in pyridine is HOO⁻ (3.0)>$t$-BuOO⁻(2.1)>O$_2$⁻·(1.0)>HO⁻[MeOH](0.3)>HO⁻[$H_2O$](0.2). For CCl$_4$ the relative reactivities are HOO⁻(1.8)>$t$-BuOO⁻(1.2)>O$_2$⁻·(1.0)>HO⁻(MeOH)(0.6)>HO⁻ ($H_2O$)(0.4), and for PhCl$_6$; HOO⁻(5.0)>$t$-BuOO⁻(2.9)>O$_2$⁻·(1.0)>HO⁻[MeOH](0.2)>HO⁻ ($H_2O$)(0.0).

Although Br(Me)CHCN has two active sites for reaction with HOO⁻ (Br and CN), its relative rate is similar to that for $n$-BuBr and it reacts with HOO⁻ at the bromo carbon to yield Br⁻, NCO⁻, and MeCH(O). In contrast, Br(Me)CHCN reacts with HO⁻ to give HO(Me)CHCN. The apparent reaction rate ($k/[S]$) between HOO⁻ and Br(Me)CHCN is 9.8 x 10$^2$ $M^{-1}$ s$^{-1}$ in MeCN, which is almost 10$^4$ times faster than the apparent rate for the reaction of HOO⁻ with MeCN ($k/[S]$ = 2.1 x 10$^{-1}$ $M^{-1}$ s$^{-1}$].[54]

The relative reactivity of HOO⁻/HO⁻[MeOH] with primary halocarbons ($n$-BuBr, PhCH$_2$Br, BrCH$_2$CH$_2$Br) is about 10 in pyridine (Table 8-5), which compares with a ratio of 13 for their reaction with BrCH$_2$C(O)OH in $H_2O$[55] and 35 for reaction with PhCH$_2$Br in 50% acetone/water.[56] Although O$_2$⁻· is a powerful nucleophile in aprotic media, it does not exhibit such reactivity in water, presumably because of its strong solvation by that medium and its rapid hydrolysis and disproportionation.[52] Kinetic and electrochemical studies for the reaction of O$_2$⁻· with primary halocarbons confirm that the initial step is rate limiting and first order with respect to substrate and O$_2$⁻·, and occurs via a nucleophilic reductive displacement of halide ion with inversion of configuration (S$_N$2).[23,24,26,45] The other oxy anions of Table 8-5 are believed to react via an analogous nucleophilic displacement.

In contrast, the reactions of CCl$_4$ with oxy nucleophiles ($YO$⁻) cannot proceed via an S$_N$2 mechanism because the carbon center is inaccessible. The primary step for the reaction of O$_2$⁻· with CCl$_4$ appears to involve a nucleophile attack of a chlorine to give Cl$_3$COO· and Cl⁻.[25] The reactions of CCl$_4$ with other oxy anions appears to be similar (e.g., CCl$_4$ + HO⁻ → Cl$_3$COH + Cl⁻).[25] However, chloroform is deprotonated by HO⁻ to give dichlorocarbene (HCCl$_3$ + HO⁻ → :CCl$_2$ + $H_2O$ + Cl⁻).[25] Thus, the relative reactivity of oxy anions with HCCl$_3$ depends on their relative Brønsted basicity ($pK_a$; HO⁻[MeOH]>HO⁻[$H_2O$]>HOO⁻ >O$_2$⁻·) rather than their nucleophilicity.

The reactions of hexachlorobenzene (PhCl$_6$) with the oxy anions are unique. Thus, HOO⁻ is five times as reactive as O$_2$⁻· and 33 times as reactive as HO⁻[MeOH] (Table 8-5). Superoxide ion reacts with PhCl$_6$ via an initial nucleophilic addition followed by displacement of Cl⁻.[47,48] The other oxy anions

Table 8-5 The Relative Reactivities of HO⁻(MeOH)/HOO⁻/O₂⁻ with Halocarbons and Esters in Pyridine [0.1 $M$ (Et₄N)ClO₄], and the Kinetics for the Reaction of O₂⁻ with Substrates

| Substrate (1–10 mM) | $[k_{YO^-}/k_{O_2^-}]^a$ | | | $O_2^-$ | $k_{O_2^-}/[S]^b$ ($M^{-1}\,s^{-1}$) |
| --- | --- | --- | --- | --- | --- |
| | $YO^- = HO^-[MeOH]^c$ | $HO^-[H_2O]$ | $HOO^{-d}$ | | |
| CCl₄ | 0.6±0.3 | 0.4±0.2 | 1.8±0.6 (1.2)$^e$ | 1.0 | (1.4±0.5) × 10³ |
| $n$-BuBr | 0.3±0.2 | 0.2±0.1 | 3.0±1.0 (2.1)$^e$ | 1.0 | (1.0±0.1) × 10³ |
| CH₂Br₂ | 0.3±0.2 | | 3.1±1.0 | 1.0 | (2.3±0.5) × 10² |
| BrCH₂CH₂Br | 0.3±0.2 | | 2.7±1.0 | 1.0 | (1.8±0.3) × 10³ |
| PhCH₂Br | 0.3±0.2 | 0.2±0.1 | 3.0±1.0 | 1.0 | >3 × 10³ |
| Br(Me)CHCN | 0.3±0.2 | | 2.8±1.0 | 1.0 | (1.4±0.3) × 10³$^f$ |
| PhCl₆ | 0.15±0.1 | 0.0 | 5.0±1.5 (2.9)$^e$ | 1.0 | (1.0±0.2) × 10³ |
| MeC(O)OPh | 0.2±0.1 | | 4.5±1.5 | 1.0 | (1.6±0.5) × 10²$^h$ |
| MeC(O)OEt | 0.2±0.15 | | 4.7±2.0 | 1.0 | (1.1±0.2) × 10⁻²$^h$ |

$^a$ Determined from the impact of $YO^-$ upon the rate of disappearance of $O_2^-$ in the presence of excess substrate (Ref. 26); rate of disappearance of $O_2^-$ monitored by linear-sweep voltammetry.
$^b$ Determined from the ratio of $i_{p,a}/i_{p,c}$ for the cyclic voltammogram of $O_2$ in DMF in the presence of excess substrate (Ref. 45).
$^c$ $HO^-$(MeOH) from (Bu₄N)OH in MeOH.
$^d$ $HOO^-$ prepared either from $O_2^-$ plus PhNHNH₂ or HOOH plus $HO^-$.
$^e$ $t$-BuOO⁻.
$^f$ $k/[S]$ = 3.5 × 10² $M^{-1}\,s^{-1}$ in MeCN.
$^g$ The primary product is Cl₅PhOMe.
$^h$ Kinetics in pyridine.

probably follow the same pathway. In the case of $HOO^-$, the displacement of $Cl^-$ gives $PhCl_5OOH$, which deprotonated by a second $HOO^-$ (or $HO^-$) to give $PhCl_5OO^-$. The latter displaces an adjacent chlorine to give the $o$-quinone.

The relative rates of reaction between esters and oxy anions $[HOO^-$ $(4.5) > O_2^{-\cdot}(1.0) > HO^-[MeOH](0.2)]$ are similar to those for $PhCl_6$, which indicates that both substrates undergo an initial nucleophilic addition to an unsaturated carbon (carbonyl and aryl-chlorine, respectively).[54] When a suitable leaving group is present, there is a net nucleophilic substitution.

The reactivity of oxy anions with nitriles relative to that for $O_2^{-\cdot}$ cannot be measured because the latter does not give a net reaction. However, the reaction between MeCN and $HOO^-$ is rapid $(k_{HOO^-}/[S], 0.21\ M^{-1}\ s^{-1})$ via nucleophilic addition to the unsaturated carbon.[54]

$$MeC{\equiv}N + HOO^- \rightarrow [(Me)(HOO)C=N^-] \rightarrow [(Me)(^-OO)C=NH] \xrightarrow{H_2O} [(Me)(HOO)(^-O)CNH_2]$$

$$\downarrow H_2O$$

$$MeC(OH)(O^-)NH_2 + HOOH$$

$$(8\text{-}27)$$

Intramolecular rearrangement and hydrolysis yield the ($HO^-$) adduct of acetamide. The $(k_{HOO^-}/k_{HO^-})$ ratio for $p$-NCPhC(O)OH in water[55] is about $10^3$ and for PhCN in 50:50 $H_2O/Me_2C(O)$[56] is $10^{4.6}$.

## Oxidative phosphorylation

The electron-transfer energetics from the neutralization of hydroxide ion by hydronium ion

$$H_3O^+ + HO^- \rightarrow 2\,H_2O \qquad (8\text{-}28)$$

is utilized in biology for the reversible storage of chemical energy. This is accomplished in a process known as *oxidative phosphorylation*, which involves a proton-induced condensation of the adenosine disphosphate protein ($ADP^{3-}$) with a phosphate ion ($P_i^{2-}$) to give adenosine triphosphate ($ATP^{4-}$)

$$ADP^{3-} + P_i^{2-} + H_3O^+ \rightleftharpoons ATP^{4-} + 2\,H_2O, \quad \Delta G, 5\ \text{kcal mol}^{-1} \qquad (8\text{-}29)$$

The energy input is accomplished by a proton gradient/flux for the oxidation of a phosphate base [$HO^-$ transfer from $HOP(O)O^-)_2$]. The oxidized product is stabilized by the nucleophilic addition of ADP to give ATP. Scheme 8-9 outlines the several steps of the energy-input and energy-output cycles in oxidative phosphorylation. The net transduction is the neutralization (via a single-electron transfer) of a hydroxide adduct [$HOP(O)(O^-)_2$] (the one-electron reductant) by a hydronium ion ($H_3O^+$)

$$H_3O^+ + (HO^-)[P(O)_2O^-] \xrightarrow{\text{SET}} 2\,H_2O + P(O)_2O^- \qquad (8\text{-}30)$$
$$[HOP(O)(O^-)_2]$$

Hence, the phosphorylation cycles represent a poised system for the reversible transfer of electrons from oxy anions $[(HO)RO^- \rightarrow (HO)RO\cdot + e^-]$ to hydronium ions $(H_3O^+ + e^- \rightarrow H\cdot + H_2O)$, which is facilitated by (1) the coupling of the respective products to form $H_2O$ $[(HO)RO\cdot + H\cdot \rightarrow H_2O + R(O);\; -\Delta G_{BF},\; 111\text{ kcal}$ $mol^{-1}]$ and (2) the nucleophilic condensation reaction $[ADP^{3-} + R(O)]$. Biological systems such as cytochrome-$c$ oxidase and Photosystem II of green-plant photosynthesis produce net proton fluxes during turnover and thereby drive oxidative phosphorylation to store 5 kcal per mole of ATP produced from one mole of hydronium ions.

*Scheme 8-9  Oxidative phosphorylation*

a.  *Acid–base chemistry of HOP(O)$_2$.*

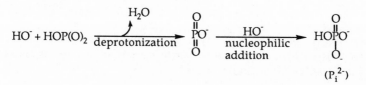

b.  *Phosphorylation.*  (i)  Energy input by $(H_3O^+)$ oxidation of $HOP(O)(O^-)_2$

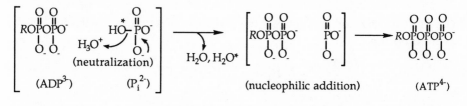

$$ADP^{3-} + H_3O^+ + P_i^{2-} \longrightarrow ATP^{4-} + H_2O + H_2O^*,\; \Delta G,\, 5\text{ kcal}$$

(ii)  Energy output by $(H_2O)_2$ reduction of $ATP^{4-}$
$$ATP^{4-} + 2\,H_2O \longrightarrow ADP^{3-} + P_i^{2-} + H_3O^+,\; -\Delta G,\, 5\text{ kcal}$$

**Summary**

1.   Solvation of $HO^-$ (and other oxy anions, $YO^-$) affects to a major degree the thermodynamics and kinetic for its reaction with electron-acceptor molecules. Solvation energies determine the ionization energy of $HO^-$ (and its redox potential as an electron donor) and the electron affinity of the electron-acceptor molecule.

2.  All reactions of oxy anions (and of nucleophiles) can be classified as a single-electron shift that leads to a polar pathway or an SET pathway.

3.  The barrier to reaction, which includes kinetic and thermodynamic components, is controlled by the solvation energies of the reactants and products, and their redox potentials and bond energies. The electron-transfer propensity of $HO^-$ increases in proportion to the covalent bond energy of the product -OH adduct. For the reaction $(YO^- + RX \rightarrow YOR + X^-)$ the nucleophilicity of $YO^-$ is proportional to the negative of its oxidation potential and to the bond energy of $YO$-$R$.

    Nucleophilicity $(-\Delta G_{reac.}) = -(E^{\circ'}YO^-/YO\cdot) 23.1 + (-\Delta G_{BF})_{YO-R}$      (8-31)

4.  Most of the known reactions of $HO^-$ that produce free radicals probably do not involve a direct single-electron transfer from $HO^-$ in the primary step because an SET primary step is usually highly endothermic; the primary step more often is an approximately thermoneutral polar reaction (polar-group transfer or polar-group coupling), with secondary reactions producing free radicals that are coupled to form stable $M$–OH bonds ($M$ is a molecule or metal atom with an unpaired electron).

5.  The relative nucleophilicity of oxy anions $(YO^-)$ in aprotic media is in the order $HOO^- > t$-$BuOO^- > O_2^{-\cdot} > HO^-(MeOH) > HO^-(H_2O)$.

### References

1.  Pross, A. *Acc. Chem. Res.* **1985**, *18*, 212.
2.  (a) Parsons, R. *Handbook of Electrochemical Constants*. Butterworths: London, 1959, pp. 69-73; (b) Bard, A.J.; Parsons, R.; Jordan, J. *Standard Potentials in Aqueous Solution*. New York: Marcel Dekker, 1985.
3.  Lide, D. R. (ed.). *CRC Handbook of Chemistry and Physics*, 71st ed. Boca Raton, FL: CRC, 1990, 9-86–98.
4.  Tsang, P. K. S.; Cofre′, P.; Sawyer, D. T. *Inorg. Chem.* **1987**, *26*, 3604.
5.  Pearson, R. G. *J. Am. Chem. Soc.* **1986**, *108*, 6109.
6.  Bohme, D. K.; Mackay, G. I. *J. Am. Chem. Soc.* **1981**, *103*, 978.
7.  Tanner, S. D.; Mackay, G. I.; Bohme, D. K. *Can. J. Chem.* **1981**, *59*, 1615.
8.  Evanseck, J. D.; Blake, J. F.; Jorgensen, W. L. *J. Am. Chem. Soc.* **1987**, *109*, 2349.
9.  Landini, D.; Maia, A. *J. Chem. Soc. Chem. Commun.* **1984**, 1041.
10. Barrette, W. C., Jr.; Johnson, H. W., Jr.; Sawyer, D. T. *Anal. Chem.* **1984**, *56*, 1890.
11. Popovych, O.; Tomkins, R. P. T. *Nonaqueous Solution Chemistry*. New York: Wiley, 1981, Chap. 10.
12. Ross, F.; Ross, A. B. (eds.). *Natl. Stand. Ref. Data Ser., Natl. Bur. Stand (U.S.)*. Washington, D. C.: U.S. Printing Office, 1977, Vol. 59.
13. Walling, C. *Acc. Chem. Res.* **1975**, *8*, 125.
14. Kunai, A.; Hata, S.; Ito, S.; Sasaki, K. *J. Am. Chem. Soc.* **1986**, *108*, 6012.
15. Roberts, Jr., J. L.; Morrison, M. M.; Sawyer, D. T. *J. Am. Chem. Soc.* **1978**, *100*, 329.

16. Sawaki, Y.; Ogada, *Bull. Chem. Soc. Jpn.* **1981**, *54*, 793.
17. Payne, G. B.; Deming, P. H.; Williams, P. H. *J. Org. Chem.* **1961**, *26*, 659.
18. Murdoch, J. R. *J. Am. Chem. Soc.* **1983**, *105*, 2159.
19. Buncel, E.; Kesmarky, S.; Symons, E. A. *J. Chem. Soc. Chem. Commun.* **1971**, 120.
20. (a) Mullay, J. *J. Am. Chem. Soc.* **1986**, *108*, 1770; (b) Jug, K.; Epiotis, N. D.; Buss, S. *J. Am. Chem. Soc.* **1986**, *108*, 3640; (c) Sanderson, R. T. *Polar Covalence.* New York: Academic Press, 1983, pp. 181, 194.
21. Sisler, H. H.; Vanderwerf, C. A. *J. Chem. Educ.* **1980**, *57*, 42.
22. (a) Blyumenfeld, L. A.; Bryukhovetskaya, L. V.; Fomin, G. V.; Shein, S. M.; *Russ. J. Phys. Chem.* **1970**, *44*, 518; (b) Chanon, M.; Tobe, M. L. *Angew. Chem. Int. Ed. Engl.* **1982**, *21*, 1; (c) Chanon, M. *Bull. Chim. Soc. France* **1982**, II-197; (d) Eberson, L. *Adv. Phys. Org. Chem.* **1982**, *18*, 79.
23. San Filipo, Jr., J.; Chern, C.-I.; Valentine, J. S. *J. Org. Chem.* **1975**, *40*, 1678.
24. Johnson, R. A.; Nidy, E. G. *J. Org. Chem.* **1975**, *40*, 1680.
25. Roberts, J. L., Jr.; Calderwood, T. S., Sawyer, D. T. *J. Am. Chem. Soc.* **1983**, *105*, 7691.
26. Roberts, J. L., Jr.; Sawyer, D. T. *J. Am. Chem. Soc.* **1981**, *103*, 712.
27. Kane, C.; Roberts, J. L., Jr., University of Redlands. Unpublished data.
28. Matsumoto, S.; Sugimoto, H.; Sawyer, D. T. *Chem. Res. Tox.* **1988**, *1*, 19.
29. Meyers, C. Y.; Malte, A. M.; Matthews, W. S. *J. Am. Chem. Soc.* **1969**, *91*, 7510.
30. Gibian, M. J.; Sawyer, D. T.; Ungermann, T.; Tangpoonpholvivat, R.; Morrison, M. M. *J. Am. Chem. Soc.* **1979**, *101*, 640.
31. Forrester, A. R.; Purushotham, V. *J. Chem. Soc. Chem. Commun.* **1984**, 1505.
32. Roberts, J. L., Jr.; Sugimoto, H.; Barrette, W.C., Jr.; Sawyer, D. T. *J. Am. Chem. Soc.* **1985**, *107*, 4556.
33. Bishop, C. A.; Tong, L. K. *Tetrahedron Lett.* **1964**, 3043.
34. Roberts, J. L., Jr.; Calderwood, T. S.; Sawyer, D. T. *J. Am. Chem. Soc.* **1984**, *106*, 4667, and references therein.
35. Goolsby, A. D.; Sawyer, D. T. *Anal. Chem.* **1968**, *40*, 1978.
36. Hercules, D. M.; Lytle, F. E. *J. Am. Chem. Soc.* **1966**, *88*, 4745.
37. Srivatsa, G. S.; Sawyer, D. T. *Inorg. Chem.* **1985**, *24*, 1732.
38. Tsang, P. K. S.; Sawyer, D. T. *Inorg. Chem.* **1990**, *29*, 2848.
39. Shin, K.; Kramer, S.K.; Goff, H.M. *Inorg. Chem.* **1987**, *26*, 4103.
40. Hojo, M.; Sawyer, D. T.; *Inorg. Chem.* **1989**, *28*, 1201.
41. Endo, T.; Miyazawa, T.; Shiihasi, S.; Okawara, M. *J. Am. Chem. Soc.* **1984**, *106*, 3877.
42. Ritchie, C. D. *Acc. Chem. Res.* **1972**, *5*, 348.
43. Shaik, S. S.; Pross, A. *J. Am. Chem. Soc.* **1982**, *104*, 2708.
44. Edwards, J. O.; Pearson, R. G. *J. Am. Chem. Soc.* **1962**, *84*, 16.
45. Merritt, M. V.; Sawyer, D. T. *J. Org. Chem.* **1970**, *35*, 2157.
46. Calderwood, T. S.; Sawyer, D. T. *J. Am. Chem. Soc.* **1984**, *106*, 7185.
47. Sugimoto, H.; Matsumoto, S.; Sawyer, D. T. *J. Am. Chem. Soc.* **1987**, *109*, 8081.
48. Sugimoto, H.; Matsumoto, S.; Sawyer, D. T. *Environ. Sci. Technol.* **1988**, *22*, 1182.
49. Magno, F.; Bontempelli, G. *J. Electroanal. Chem.* **1976**, *68*, 337.
50. Gibian, M. J.; Sawyer, D. T.; Undermann, T.; Tangpoonpholvivat, G.; Morrison, M. M. *J. Am. Chem. Soc.* **1979**, *101*, 640.
51. San Fillipo, J. Jr.; Romano, L. J.; Chern, C.-I.; Valentine, J. S. *J. Org. Chem.* **1976**, *41*, 586.
52. Sawyer, D. T.; Valentine, J. S. *Acc. Chem. Res.* **1981**, *14*, 393.

53. Cofré, P.; Sawyer, D. T. *Inorg. Chem.* **1986**, 25, 2089.
54. Tsang, P. K. S.; Jeon, S.; Sawyer, D. T. *Inorg. Chem.* **1991**, submitted.
55. McIssac, J. E., Jr.; Subbaraman, L. R.; Subbaraman, J.; Hulhausen, H. A.; Behrman, E. J. *J. Org. Chem.* **1972**, 37, 1037.
56. Pearson, R. G.; Edgington, D. N. *J. Am. Chem. Soc.* **1962**, 84, 4607.

# INDEX

215